Wakefield Press

ALTERNATIVE INTERVENTIONS

Alan Mayne holds a ResearchSA Chair at the University of South Australia, where he is Professor of Social History in the David Unaipon College of Indigenous Education and Research, and the Hawke Research Institute. He holds a PhD (1980) from the Australian National University and worked until 2005 at the University of Melbourne. He has also been a Woodrow Wilson Fellow in Washington DC, a Senior Fulbright scholar at Boston and Berkeley, and a visiting Professorial Fellow at Jawaharlal Nehru University in New Delhi. His core interests relate to social equity and sustainability.

By the same author

Fever, Squalor and Vice
(University of Queensland Press, 1982)
The Imagined Slum
(Leicester University Press, 1993)
Reluctant Italians?
(Dante Alighieri Society, Melbourne, 1997)
The Archaeology of Urban Landscapes
(with Tim Murray, Cambridge University Press, 2001)
Hill End: An Historic Australian Goldfields Landscape
(Melbourne University Press, 2003)
Eureka: Reappraising an Australian Legend
(Network Books, 2006)
Beyond the Black Stump: Histories of Outback Australia
(Wakefield Press, 2008)
Building the Village: A History of Bendigo Bank
(Wakefield Press, 2008)
Outside Country: Histories of Inland Australia
(with Stephen Atkinson, Wakefield Press, 2011)

ALTERNATIVE INTERVENTIONS

Aboriginal Homelands, Outback Australia and the Centre for Appropriate Technology

ALAN MAYNE

Wakefield Press
1 The Parade West
Kent Town
South Australia 5067
www.wakefieldpress.com.au

First published 2014

Cover design by Michael Deves, Wakefield Press
Typeset by Wakefield Press
Printed in Australia by Lane Print and Post

National Library of Australia Cataloguing-in-Publication entry

Author:	Mayne, Alan, author.
Title:	Alternative interventions: Aboriginal homelands, outback Australia and the Centre for Appropriate Technology / Alan Mayne.
ISBN:	978 1 74305 272 3 (paperback).
Notes:	Includes index.
Subjects:	Centre for Appropriate Technology (Alice Springs, N.T.)
	Research institutes – Australia – History.
	Appropriate technology – Northern Territory.
	Community development – Northern Territory.
	Aboriginal Australians – Northern Territory – Economic conditions.
	Aboriginal Australians – Northern Territory – Social conditions.
Dewey Number:	338.927

Contents

Illustrations

All the illustrations in this book were prepared from originals belonging to CAT. The illustration captions were provided by CAT.

Acknowledgements

Research for this history was generously supported by the Centre for Appropriate Technology (CAT), which gave me unrestricted access to its archives, helped to arrange interviews, and provided useful illustrations. However, *Alternative Interventions* is not a commissioned history and my interpretations should not necessarily be identified with those of CAT itself.

The book is the result of a conversation between two distinguished Australians. One of them was Basil Hetzel, a former Chancellor of the University of South Australia and a world-renowned medical scientist. The other was Bruce Walker, founder and director of CAT since 1980. Basil suggested to Bruce that I record CAT's history, and Bruce and the CAT board agreed. I readily accepted because I could see that CAT's history was significant and deserved general recognition. I have had many conversations with Basil and Bruce, which have not only enriched my understanding of CAT's achievements, but have also confirmed my respect for these two 'rocks of Gibraltar' in whitefella advocacy of rights and livelihoods for Aboriginal and Torres Strait Islander Australians.

However, there is a third presence behind my development of this book, which sustained me during the long lag between beginning and completing it. That presence is Jim Bray. Jim was the foundation chairman of CAT's Aboriginal board of management from 1990 until 2010. Jim and I yarned often when I visited Alice Springs, because we initially shared adjoining apartments in CAT's former Priest Street headquarters. Jim and I share an affection for Adelaide, where Jim (an Eastern Arrernte man) went to school from the late 1940s: he was one of that iconic cohort of Aboriginal students from Alice Springs who went to St Francis House and thereafter carved out remarkable careers. We also share affection for the MacDonnell Ranges, which Jim paints in impressive acrylics. Many of those who have dealt with Jim over the years think of him as a warrior who takes few prisoners, but his essence is compassion and inclusiveness, albeit conditioned by tireless advocacy of Aboriginal Australian futures.

I am grateful to the many CAT staff and affiliates, past and present, who gave information and advice, and who agreed to be interviewed. Among these I highlight here the input of Peter Taylor, who, sadly, passed away before the book's completion. He did, however, witness a preliminary version of CAT's history, based upon my initial research and written by Glenn Morrison, which was published as a special issue of CAT's *Our Place* Magazine in April 2010. Jim

Bray, Kurt Seemann, Bruce Walker and Metta Young read and commented on my original text.

Paul Wallace skilfully edited and indexed my narrative, and prepared the maps. Walker and Young selected the illustrations. Wakefield Press crafted this assemblage into a book. Judy King shared the highs and lows of researching and writing it, smoothing out my lapses in logic and expression. I especially thank her, my dear companion.

Chapter 1

Introduction

This book explores the history of the Centre for Appropriate Technology (CAT), an Aboriginal governed organisation that began, and is still based, in Alice Springs at the centre of the Australian Outback. CAT applies science and technology to enhance Aboriginal livelihoods and underpin community wellbeing throughout the Outback. It seeks to transform the lives of Aboriginal Australians, which when measured by most social indicators are far more constrained than those enjoyed by the majority of non-Aboriginal Australians.[1] CAT was established in 1980, and has become a large, national, not-for-profit corporation with an annual budget in excess of $25 million. It provides information and practical assistance to Outback communities for housing, water supply, energy, waste, telecommunications, transport and other infrastructure needs. It offers education and training programmes, community planning and project management. It does so by tailoring its services to fit the socio-economic development and cultural priorities of the communities it partners with. Understanding CAT's history can usefully inform future community choices and government policy emphases in Outback Australia. As CAT's founder, Bruce Walker, urged, 'We should look back, learn and move forward'.[2] Peter Taylor, who succeeded Walker as CAT's Chief Executive Officer in 2010, cautioned that without a sense of history 'it is very easy to forget what has worked and what has failed in Indigenous affairs; as an area of public policy it is prone to a lot of forgetting'.[3]

As Taylor's words imply, CAT's history is best considered alongside the evolving policy approaches of the Australian, state and Territory governments for Aboriginal people. This book tells the story of CAT's 'bottom-up' work in small Aboriginal communities to apply 'appropriate technology' (a term explained below), rather than dwelling on the 'top-down' approaches adopted by government and other mainstream agencies. CAT often found itself 'push[ing] the boundaries to the edge'[4] of what governments would countenance in

Aboriginal affairs, increasingly so as neoliberal thinking came to dominate policy during the 1990s and the early 21st century, but also during the 1970s and 1980s, as human rights, self-determination and self-management principles replaced earlier social engineering policies that had been designed to assimilate Aboriginal people into mainstream Australia. CAT challenged prevailing public opinion of Aboriginal people: stereotypes about a monolithic Stone-Age culture that stood remote from the modern world; of social marginalisation from mainstream values and normal behaviour; and about the unsuitability of Aboriginal Australians for permanent jobs in the 'real' economy. Throughout its history, CAT has sought to enhance Aboriginal wellbeing by employing alternative strategies to those that flow from the conventional assumption that Aboriginal '"difference" is a "deficit" to be remedied, rather than a "choice" to be respected'.[5] The Australian government's latest programme for 'Closing the Gap on Indigenous Disadvantage', begun in 2009, leaves CAT's Aboriginal leaders unimpressed. 'Bridging the Gap!' snorts Noel Hayes, a Kaytej man from Barrow Creek, a CAT board member and former regional chairman for the Aboriginal and Torres Strait Islander Commission (ATSIC). He adds: 'famous words now ... They'll never bridge the gap, mate!'[6] His meaning is clear: nothing will change, so long as decision-making opportunities continue to be drained out of Aboriginal communities in the interests of top-down reform agendas.

CAT champions the livelihoods of Aboriginal people living in what mainstream Australia calls the 'remote' regions of the Outback. CAT's experimentation with appropriate technology began here in 1980 with a hand pump designed to deliver water to the outstation communities that were resettling traditional homelands in the Western Desert. CAT's activities subsequently extended in a wide arc through the desert regions of the Northern Territory, Western Australia, Queensland and South Australia, and to the sea-gazing communities of the Gulf of Carpentaria and the tropical communities of Cape York Peninsula.

It is unambiguously the case that within this Outback zone, 'remote Aboriginal settlements operate in an extreme economic context, arising from limited economic opportunities, the small size of settlements, large distances between settlements, a lack of institutional capital, and high levels of mobility between and within settlements'.[7] However, characterisations of these settlements as 'remote' (and, by implication, marginal and dysfunctional) do not help address these constraints; still less harness the energies of Aboriginal communities for the future development of Outback Australia. Objecting to the characterisation 'remote', Walker queried, 'remote from what?' Answering

Australia's 'remote' zone and areas of CAT operations.

his own question, he suggested: 'if you live in Kintore in the far west of the Northern Territory or a community near Fitzroy Crossing in Western Australia then the remote communities are Darwin or Perth or Canberra'.[8] CAT's experiences since 1980 suggest that the main problem with remoteness in Outback Australia is that the policy-makers in the capital cities are remote from local realities, not that the communities are dysfunctional and unsustainable, nor monolithically and unchangingly 'traditional' in their cultural aspirations and lifestyles. By the Australian Outback and the remote communities within it, this book means 'that part of the [Australian] landmass that is at distance from centres of economic and political decision making'.[9] Less than 15 per cent of Australia's population (but some 25 per cent of Aboriginal Australians) live in this remote region, whereas over 85 per cent of Australians live inside a 50 kilometre wide zone around the coastline. Decisions about society, culture and the national economy are made overwhelmingly within that coastal zone, and they have largely been at the expense of remote Australia.[10]

The word 'intervention' is therefore heavily loaded with meaning when applied in Outback and Aboriginal settings. The Australian nation is itself the creation of European intervention and colonialism, and its political economy is dominated by its coastal crescent. The genesis of today's Aboriginal affairs policies was paternalistic intervention designed to protect and separate Aboriginal people from Europeans, and then to assimilate them into broader Australian society. These strategies were abandoned in the 1970s, but subsequent state interventions on behalf of Aboriginal Australians have been seen as another form of colonialism whereby 'Aboriginal people nominally enjoy the right to run their own affairs, but actually find themselves having to learn to do so according to the forms of land tenure and administrative process created for them by the state'.[11] Paradoxically, state interventions in the name of Aboriginal self-determination and self-help have increased Aboriginal dependence. This is not a uniquely Australian experience. As E.F. Schumacher warned, 'Unintentional neocolonialism is far more insidious and infinitely more difficult to combat than neocolonialism intentionally pursued. It results from the mere drift of things, supported by the best intentions'.[12]

This problem was highlighted in Australia during 2007 when the Australian government, responding to allegations of the widespread sexual abuse of Aboriginal children, launched the Northern Territory Emergency Response (often called the Northern Territory Intervention), directly overseeing remote Aboriginal communities in what Aboriginal leader Patrick Dodson called a reassertion of the old 'assimilation agenda'.[13] The Intervention suspended the operation of the *Racial Discrimination Act 1975* in the Territory's Aboriginal communities. Rose Kunoth-Monks, a spokeswoman for the Utopia homelands community and a member of CAT's board, described the Intervention as 'a huge thing. It was assault. Assault to such an extent that it traumatised all of us'.[14] CAT's alternative interventions in Outback Australia instead sought, and still seek, to partner with Aboriginal communities, drawing upon, instructing and empowering local knowledge and action. CAT uses appropriate technology as the instrument of these positive interventions.

****

There are other well-known CATs in the world. The best known of these, which predates Australia's CAT, was established in Wales in 1973. It is called the Centre for Alternative Technology (CAT). Environmentalists and sustainable living activists in Britain popularised the term 'alternative technology' during the 1970s to refer to technologies that were more environmentally friendly than conventional technologies.[15] The Welsh centre began as 'a community dedicated to eco-friendly principles', and evolved into a large-scale education facility and visitor centre that advocates sustainable 'green living': environmentally sensitive building construction, renewable energy and energy efficiency, eco-sanitation, woodland management and organic gardening and farming.[16] The Welsh centre in turn supported the beginnings of another CAT, the Urban Centre for Appropriate Technology, which was established in Bristol in 1979 and subsequently became the Centre for Sustainable Energy (CSE).[17] However, Australia's CAT is closer in both name and concept to the National Center for Appropriate Technology (NCAT) in the United States, which was established in Butte, Montana, in 1976, initially with generous federal funding. NCAT's aim is to support 'economically disadvantaged people by providing information and access to appropriate technologies that can help improve their lives'.[18]

NCAT and Australia's CAT, unlike CAT and CSE in Britain, use the phrase 'appropriate technology'. This describes technological applications that are not only sensitive to the environment, but are also appropriate to the social and cultural contexts of their use. As Dale Jones (an Aboriginal woman who worked as an engineer for mining giant Rio Tinto in Western Australia, and became the youngest member of CAT's board in 2008) remarked to me, 'technology is not necessarily the high end of things, computers and complicated things. Technology is what helps to make your life easier'.[19] The appropriate technology movement aims to achieve sustainable and environmentally responsible human development through outcomes that empower and benefit local communities. It pursues social equity goals to support disadvantaged neighbourhoods within the 'developed world', and livelihood options for households and communities in the 'developing world'.

Advocates of appropriate technology drew upon Mahatma Gandhi's writings against British colonialism in India, and his advocacy of a new social order for post-colonial India. His *Hind Swaraj*, first published in English in 1938, scathingly attacked the large-scale technological foundations of Western civilisation. His *Constructive Programme*, released in 1941, proposed an alternative developmental pathway for India through decentralised production in villages. 'The greater the decentralization of labour', Gandhi predicted, 'the simpler and cheaper the tools' that communities would require. The result, he said, would be a self-sufficient and equitable India: 'When we have become village-minded, we will not want imitations of the West or machine-made products, but we will develop a true national taste in keeping with the vision of a new India in which pauperism, starvation and idleness will be unknown'.[20]

The phrase 'appropriate technology' was popularised during the 1960s and 1970s by Schumacher, an admirer of Gandhi and adviser to post-independence India. Schumacher and his collaborators toyed with the term 'intermediate technology' during the 1960s but, in his widely influential book

Small is Beautiful (1973), Schumacher used the term 'appropriate technology' and defined it as 'technology with a human face'.[21] Schumacher argued that simple, small-scale, and resilient technologies could 'open … up avenues of constructive action' for disadvantaged social groups by providing them with reliable means to deliver what they so conspicuously lacked: sustainable livelihoods.[22] Appropriate technology could thus underpin 'a new life-style, with new methods of production and new patterns of consumption: a life-style designed for permanence'.[23] In Schumacher's eyes, appropriate technology was a form of 'self-help technology, or democratic or people's technology' to be used in support of equitable and sustainable community development.[24] Appropriate technology organisations mushroomed around the world during the 1970s and, by 1980 when CAT began in Alice Springs, there were over 1,000 such organisations worldwide.

The majority of these appropriate technology organisations pursued community development objectives in the 'developing world'. Although community development has been criticised by some as yet another unhelpful intervention in the affairs of Aboriginal Australians, this view, while valid in some respects, is limited and ungenerous. The international community development movement was sustained in part by philanthropic organisations such as the Ford Foundation in the aftermath of the post-1945 disintegration of Europe's colonial empires. It was boosted during the 1950s and 1960s by post-independence India's emphasis on rural community development schemes under the prime ministership of Gandhi's former disciple, Jawaharlal Nehru. India's activities were widely copied by other post-colonial nations. During the late twentieth century, community development activities were promoted by international aid agencies, endorsed by the World Bank, and further reinforced by the United Nations' Millennium Development Goals, announced in 2000, which were designed 'to free … humanity from the shackles of extreme poverty, hunger, illiteracy and disease'.[25]

CAT Australia's activities were directly influenced by the international community development movement, and key figures within CAT have worked on international community development projects. CAT's founder, Bruce Walker, undertook consultancies in Burma, Kenya, Tanzania and Vanuatu. Steve Fisher, CAT's deputy director during the early 2000s, was previously a senior executive in the international community development agency Practical Action, and subsequently became a director of Community Works. Mark Moran, who led CAT's Queensland operations during the 1990s and returned as its research manager for several years in the new century, also undertook community development work in China, Bolivia, East Timor and Lesotho,

Bruce Walker in Laukkai, Eastern Shan State Burma near the China Border working with the United Nations Capital Development Fund (UNCDF) Village Water Supply Project, 1991.

and later became a senior official in World Vision Australia. Doug Porter, who carried out a major review of CAT on behalf of ATSIC during the 1990s, and subsequently co-authored with Walker a series of community development reports on the Northern Territory, had previously worked for the Freedom from Hunger organisation, and later became senior governance adviser for the Asian Development Bank and the World Bank.[26] CAT contracted with international architecture firm, Hassell, to supply staff members Moran, Michael Adams and Bob Lloyd for projects in China and Mongolia. Richard Callahan, a 14 year veteran of CAT's extension services, was recruited by the Red Cross to develop gravity-fed water supplies in Tibet.

The history of community development activities reveals many examples of inappropriate 'top-down' modernising interventions in the local affairs of communities in the 'developing world' and in remote Australia.[27] By contrast, CAT's achievement in Outback Australia has been to add the dimension of cultural respect and two-way knowledge generation to the design and implementation of appropriate technologies, so as to ground community development expertise in local knowledge systems and decision making, in order to extend Schumacher's goal of sustainable livelihoods to Aboriginal

Dr Mark Moran working on the Qinghai Poverty Alleviation Project in China April, 1993.

and Torres Strait Islander people. As Moran astutely remarked, CAT's great strength was its willingness to develop regional appropriate technology solutions in the 'intercultural space' that overlapped the mindsets of local communities and development experts.[28] CAT defined appropriate technology as the 'range of skills, equipment and techniques which meet [the] particular economic, cultural, environmental and social needs of people'.[29] By contrast, 'Technologies developed in Western cultures carry with them a set of assumptions and values which conflict directly with the values and lifestyle of Aboriginal people living in remote communities', and which obscure 'awareness of Aboriginal science and technology'.[30] The organisation operated from the premise that for a technology to work long-term in the Outback it 'has to be appropriate to the social, cultural and physical environment into which it is introduced. Systems are also more likely to function longer if they are within the capacity of consumers to operate, maintain, and afford'.[31] It followed that technological needs and livelihood solutions could not simply be imposed on Outback communities; they needed to be chosen by the participating communities and

A typical village reception for the Australian International Development Assistance Bureau (AIDAB) team reviewing the Burma Village Water Supply Project in the Dry Zone of Burma, 1983.

tailored to their local conditions. Indeed CAT spent considerable time observing how Aboriginal people appropriated technology to their own ends.[32]

These were not easy lessons to learn. When in 1984 one hard-working and well-intentioned non-Aboriginal CAT worker arrived at Haasts Bluff, a long and dusty drive from Alice Springs into the West MacDonnell Ranges, to install two appropriate technology toilets, he was appalled that the local community had not prepared the site in any way: 'nothing at all had been done, not a thing … so … I just sat down and I said "if you're not going to work I'm not going to work, if you don't work I'll go back to Alice Springs"'. Several teenagers agreed to help, but 'the men of the camps just stood around and watched'. He mused, 'they think that us white fellas from Alice Springs come out there and manufacture everything, it just happens and just falls out of the sky'.[33] But he should instead have pondered a bigger issue: why was the local community so disengaged? Had their support been previously sought, or merely assumed? Walker quickly realised that appropriate technology was ultimately less about providing things like dependable toilets and washing machines to Outback communities than it was about 'a process of reconciliation of values'.[34] As he explained, 'appropriate technology is about getting things done in a way that empowers you in the short and long term. Appropriate technology has a human outcome or benefit associated with it and is usually responsive to local

Richard Callahan, a 14-year veteran of CAT's extension services at the opening of a Red Cross gravity-fed water supply in Tibet.

situations'.[35] CAT's interventions in Outback Australia were thus very different from those imposed by the Australian state. As one CAT insider remarked, 'I mean, once you get into this intervention mode and go in there and do stuff to people because you've decided it's the best thing for them, well (laughs), you know, it's not rocket science [to realise that] people are going to stand back'.[36]

CAT's early history was thus conditioned by its responsiveness to local situations and aspirations, and as the organisation matured it increasingly adopted a holistic viewpoint – reflecting the values of its Aboriginal board of management – that recognised the livelihood needs not only of discrete communities, but of the entire Outback region. Both the small- and broad-scale activities were anchored in 'CAT's vision ... of happy and safe communities of Indigenous people', and its commitment 'to secure sustainable livelihoods through appropriate technology'.[37] CAT defined sustainable livelihoods as the 'range of activities that support improved well-being through work, enterprise and trading and that can be maintained into the future … It builds upon the assets that people already have in aspiring for an improved livelihood outcome'.[38] These activities were largely pursued 'on country' in the traditional lands of Aboriginal and Torres Strait Islander people.

* * * * * * * * * * * *

CAT began as the Aboriginal homelands movement gathered momentum in Australia during the 1980s, the organisation's rapid growth being a measure of its success in providing practical support for new outstation communities. CAT's activities are still directed in large part at small communities within the homelands of Outback Australia. During the period of CAT's existence, over 20 per cent of the Australian landmass (including roughly half of the Northern Territory) has been returned to Aboriginal Australians through native title claims and land acquisition programmes, and over a thousand Aboriginal communities have been established in Outback Australia.

The homelands movement began during the late 1960s when a scattering of outstations was established on traditional lands. The movement strengthened during the 1970s after the election of the reformist Whitlam Federal Labor government in 1972. Whitlam appointed the Woodward Royal Commission into Aboriginal land rights, which led to the *Aboriginal Land Rights (Northern Territory) Act 1976* – the Northern Territory then being administered by the federal government – and the subsequent opening up for all Aboriginal Australians of land rights entitlements and access to social security benefits. Thus, from the early 1970s, outstation groups could apply to the new federal Department of Aboriginal Affairs (DAA, established late in 1972) for funding to provide basic water, shelter and communication infrastructure, and from the late 1970s the federal government's Community Development Employment Projects Scheme (CDEP), which was further expanded in the late 1980s, provided income support for local community development activities.

Outstations (settlements with fewer than 50 inhabitants) are currently home to some 22,000 people (including one quarter of the Aboriginal population of the Northern Territory), and another 100,000 people live in homelands communities of fewer than 100 inhabitants in the Northern Territory, Western Australia, Queensland and South Australia.[39] These numbers are very fluid; the essence of homelands communities is that they comprise 'one level in a hierarchy of settlements of different sizes' in which people reside, and between which they regularly move.[40] Although homelands populations are statistically insignificant at a national level, the homelands have deep cultural significance for Aboriginal Australians. As the distinguished Australian public intellectual and inaugural chairman of the federal government's Council for Aboriginal Affairs, H.C. Coombs, argued when the homelands movement was still in its infancy, the move to country was 'an attempt by Aboriginal people to ... evolve a life-style which combines what they wish to retain of the Aboriginal way with access to those goods and services from the white society on which they have come to depend'.[41] Anthropologist Jon Altman contends that

> Outstations/homelands are at the centre of Aboriginal economic, cultural and spiritual life across much of Australia ... [providing] a social setting within which Indigenous languages, ecological knowledge, culture and law can remain strong and relevant, and so underpin community development, economic initiatives and sustainable land and sea management.[42]

However, government support for small Aboriginal Outback communities rested on a fragile liberal consensus that, having begun in the 1970s, began to unravel during the 1990s. It was based on the concept of Aboriginal self-determination (the beginnings of which can be traced to the McMahon Liberal-Country Party coalition government in 1971) and universal human rights principles (expressed in the Racial Discrimination Act, the establishment of the Australian Human Rights Commission in 1986 and Australia's delayed endorsement of the United Nations Declaration on the Rights on Indigenous Peoples in 2009). Concerns about escalating and ad hoc funding of homelands communities led ATSIC (which replaced DAA in 1990) in 1996 to declare a moratorium on further outstation funding while it reviewed the existing funding mechanisms. ATSIC's National Homelands Policy (to which CAT contributed significantly) proposed minimum criteria below which settlements would not receive government assistance, and tapering levels of support from large to small populations. It was never fully implemented. ATSIC was scrapped by the Howard Liberal-National Party coalition government in 2003, and its oversight of Aboriginal affairs was divided between mainstream government departments. As neoliberalism reshaped the policy agendas of the major political parties, the Howard government became increasingly sceptical about Aboriginal homelands, and government bureaucrats gradually tightened the support that was available to them. The federal Labor governments that replaced Howard after 2007 were no more sympathetic. CDEP was effectively withdrawn from the larger regional communities in 2009, and then from the homelands in 2011, making outstation residents dependent on mainstream support services that were only available from the larger towns. As Brian Singleton, an Aboriginal Project Officer in CAT's Cairns office, recalls, when CDEP had been available to the outstations 'things were happening; it wasn't big-scale things … it was teaching their kids culture, maintaining … their homelands, but now that's all gone', with the outstations now merely 'a place where they just get out there for the day … it's sad'.[43] A key Memorandum of Understanding signed between the federal and Northern Territory governments in 2007 stated that no more funding would be provided 'to construct housing on outstations / homelands'.[44] The Australian and Northern Territory governments announced in the following year that support services for remote communities would be channelled through a network of larger urban hubs. This bilateral approach was strengthened in 2012 with the launch of the 'Stronger Futures in the Northern Territory' programme.

This developing policy approach was at odds with CAT's alternative interventions in the Outback since 1980. In 2006, criticising Indigenous Affairs Minister Amanda Vanstone, CAT argued that 'to simply declare communities are unviable and infer all should move to larger centres of economic opportunity is as much head in the sand as the expectation that people can rely solely on traditional law and culture to see them through the next twenty years'.[45] Vanstone had, late in the previous year, stated bluntly that remote Aboriginal outstations were 'cultural museums' rather than 'viable' communities.[46] Notwithstanding the conventional wisdom of non-Aboriginal Australians as to 'what they think Aboriginal lifestyle or tradition ought to be', living on country was not an uncomplicated 'attempt to resume an age-old way of life'.[47] Aboriginal self-determination did not mean a return to the Stone Age, nor to the bottling up of the Outback's growing Aboriginal populations into only the smallest and most basic of settlements. The school principal at Walungurru (a large Aboriginal community to the far west of Alice Springs that CAT had helped to establish as an outstation in 1981) remarked in 2001 that 'whether or not it suits the dominant society's aspirations for them, families are choosing to watch "action" videos, rather than sitting dutifully around their campfires recounting Dreamtime stories'.[48]

DAA fieldworkers had recognised in the early 1980s that the 'movement to outstations does not mean a return to a traditional existence', and they concluded that 'it is therefore necessary to introduce some modern technology to achieve ends which are no longer achievable by traditional means'.[49] CAT concurred with DAA on outstations and traditionalism, but fundamentally disagreed with DAA over the unmediated introduction of modern technology. CAT, and other appropriate technology organisations around the world, have been criticised for attempting to 'dumb down' advanced technologies to make them appropriate for use by 'primitive' people. This is a spurious argument in the context of Australia's CAT. Tim Rowse once suggested that a fundamental flaw in Australian government interventions in Aboriginal and Torres Strait Islander affairs since the 1970s had been 'the way Indigenous "choice" is imagined. It is not necessary to suppose that Indigenous Australians are simply driven by the inexorable logic of "modernity" (in one perspective) or by the noble imperatives of "tradition" (in the alternative perspective)'.[50] CAT filled that gap in the imagination that Rowse identified. It did not attempt to scale down modern technologies to make them fit a traditional lifestyle played out in obscure places; it sought instead to harness the knowledge, aspirations and energy of Outback communities so as to match technologies, habitats, livelihoods and Aboriginal wellbeing. As Walker reflects: 'time moves on. We weren't trying to provide timeless solutions for people that locked them in. We were trying to provide solutions that enabled people to get on with the next bit of their lives at that point in time'.[51]

CAT, through its alternative interventions throughout the Australian Outback since 1980, has sought to demonstrate that 'technology and science hold the seeds for Indigenous development and economic independence'.[52] CAT's Executive Officer Jenny Kroker, an Eastern Arrernte woman, snorts with frustration at the negative stereotypes that continue to hold sway in mainstream Australia about remote Aboriginal communities, and that purport to explain the necessity for the Northern Territory Intervention. She exclaims, 'hey you government mob, get down here and get involved, and look at what's happening on the ground! [T]here's lots of positive things at ground level; learn from those positive things and grow from them!'[53] Would that the government mob came to Outback Australia to heed Kroker's words.

Chapter 2

Beginnings: the 1980s

CAT's beginnings stem from the appointment of Bruce Walker in 1980 as Lecturer in Appropriate Technology at the Community College of Central Australia (CCCA) in Alice Springs.[1] Walker immediately set about creating a centre for appropriate technology that, as it expanded during the 1980s, acted increasingly independently of the college. By the end of the decade it had become an incorporated body with an Aboriginal board of management, and functioned as a quasi-autonomous organisation. However, neither the initial concept of a centre nor its name originated with Walker. They derive instead from Jim Pearse, the college's principal during the late 1970s, whose ideas were in turn influenced by a loose network of Aboriginal organisations, bureaucrats, technical officers and scientists working to support the Aboriginal homelands movement.

Walker's achievement during the 1980s was to translate a loose concept into reality. As he noted to a friend in the mid 1980s, 'The Centre for Appropriate Technology has mushroomed much faster than I would have dreamed'.[2] By the decade's end he could justifiably call CAT 'the foremost institution in Appropriate Technology in Australia'.[3] Walker's originality in overseeing this transformation was twofold. Firstly, he focused the fledgling centre's activities upon providing practical support for Aboriginal homelands in Central Australia, collaborating with outstation communities in an ever-widening arc around Alice Springs, and researching and developing robust technologies suited to outstation needs and conditions. Secondly, during the 1980s, as the centre's range of outstation products and associated support activities broadened and the region serviced by it expanded, Walker took on the additional role of providing technical education and training activities for remote communities. A training workshop was established in 1983, field teams commenced installation and on-site training in the homelands during 1984, and the Aboriginal Technical Worker Programme (ATWORK) began to be developed as a national training initiative from 1986.

CAT's activities helped to fill a vacuum in community development initiatives for Aboriginal people in Outback Australia during the 1980s, in the absence of any deep engagement by federal or state governments with the grassroots issues facing remote communities, and the antagonistic relationship that was developing in central Australia between the new Northern Territory government and Aboriginal groups over governance and service delivery. However, CAT necessarily had to learn the complicated new sets of rules that were being devised by governments for 'players in the funding game' that was coming to dictate Aboriginal futures in the post-assimilation period.[4] CAT's successes therefore came at a cost: its apparent independence as it loosened its ties with CCCA, and Walker's initially cautious critiques of the effects of government programmes upon remote communities, were both constrained by the shifting goalposts of government policy. Like other Aboriginal organisations that proliferated following the policy shift from assimilation to self-determination during the early 1970s, CAT – dependent upon government funding schemes to translate new ideas into practice – became entangled in a suffocating relationship with the Australian state. Charles Perkins, a leader in the post 1972 'rise of the Indigenous Sector', later conceded that 'since the early 1980s Aboriginal organizations have become preoccupied with following agendas established by others'.[5] One might 'be cheeky' and attempt to rock the boat, as Jim Bray, the first chairman of CAT's Aboriginal board of management, later recalled; but ultimately one had to toe the line if government funding was to be maintained.[6] This dependence upon state arbiters resulted in a disturbing paradox: that 'Aboriginal people [might] nominally enjoy the right to run their own affairs, but actually find themselves having to learn to do so according to the forms of land tenure and administrative process created for them by the state'.[7] This limited the choice of lifestyle and livelihood options available to Aboriginal communities during the 1980s and constrained their take-up of appropriate technology and community development opportunities.

1. Origins

In May 1979 the Alice Springs Community College (ASCC) advertised two new lectureships in Aboriginal community development. One was to be a 'Community Management Specialist' and the other an 'Appropriate Technologist'. The first advertised position was politically savvy in its alignment with evolving federal government policy emphases in Aboriginal affairs, as the Whitlam government's focus during the early 1970s on self-determination for Aboriginal communities translated into the Fraser government's interest by the late 1970s in the sustainable development of remote communities through local accountability and self-management. Thus, the new lectureship was intended to 'initiate and develop a College program in conjunction with Central Australian Aboriginal communities' in order to assist 'the administration of Aboriginal communities as they engage increasingly in self-management'. The advertisement for the second position similarly echoed the Fraser government's priorities of assisting 'individuals and communities to move through thresholds of knowledge and skills towards adequate self-management and self-reliance'. However, the core concept of 'alternative technology' would have seemed to many to have come out of nowhere.[8]

The Appropriate Technologist would 'need a background indicating an innovative capacity in one or more of architecture, building science, the several branches of engineering, technological science, agricultural or horticultural science'. A detailed position description explained that the lectureship was intended to support 'the College's program in Aboriginal Community Development', but the appointee would also establish training programmes for 'technologies ... which are appropriate to the requirements of Aboriginal Communities, groups and individuals', and 'review on a continuous basis the technologies and their appropriateness for the requirements of Aboriginal Communities in Central Australia'.[9] 'Community Development' was a well-rehearsed concept that the developed world had sought to apply in the developing world, especially in India and Africa, since the 1950s. Its well-intentioned efforts were often undermined by paternalistic mindsets that made it difficult for community development programmes to partner effectively with the Aboriginal communities they were intended to help. In Alice Springs, however, a more culturally sensitive and inclusive approach to Aboriginal community development had been promoted during the 1970s by Uniting Church social activist the Reverend Jim Downing, and his approach influenced CCCA's thinking on the scope of its new lectureships. The advertisement's emphasis upon alternative technology, echoing Schumacher's *Small is Beautiful* (1973), extended this thinking. By combining community development with appropriate technology, the college's advertisement seemed to recognise the need to match Outback technology delivery and technical training to the cultural emphases and social needs of Aboriginal communities.

Only one of the positions, that of the Appropriate Technologist, would ultimately be filled. The successful applicant was Bruce Walker, who hailed from Sydney where he had for several years been working as a schoolteacher in engineering science and woodworking. Basil Hetzel, at that time the CSIRO's divisional chief in Adelaide and an early advocate of better community development approaches for Aboriginal Australians, chuckled later that Walker

Walker with the Yolngu men who worked felling trees for the Gapuwiyak sawmill, 1973.

'just bobbed up, we were looking for somebody, and there he was, with a PhD on Arnhem Land!'[10] Walker recalls that he 'talked [his] way into the job' during an interview that stretched over three hours, although Pearse, who conducted the interview, remembers that he was in fact the stand-out applicant.[11] Walker, aged 30, held an undergraduate science degree and a PhD (1976) from the University of New South Wales, where he had excelled in the innovative multi-disciplinary programme in industrial arts. It was in part this academic background in design that had introduced him to appropriate technology, rather than any direct affinity with Schumacher. However, in equal measure, Walker's contacts in the Uniting Church steered him towards applying his academic interests in appropriate technology to the social justice activities that Schumacher had emphasised in India, and which in Australia translated readily into a commitment to maximising Aboriginal wellbeing. This interest, first expressed in his undergraduate honours research thesis at the University of New South Wales, intensified during his PhD studies when he visited the Centre of Ekistics (the science of human settlements) in Athens. From its founder, Constantinos Doxiadis, he matured his interest 'in the construction of appropriate technological hardware designed to solve problems unique to remote Aboriginal lifestyles',[12] and applied it during his two years' PhD fieldwork with the Gapuwiyak (Lake Evella) community in Arnhem Land.

Walker did not take up his appointment in Alice Springs until May 1980, almost a year after the lectureship had been advertised. Complex sets of circumstances had influenced the College to advertise for an appropriate technologist in the first place, and these were also responsible for the delay in putting the appointment into place.

* * * * * * * * * * * * * * *

The new position was not funded by the college itself, nor (after its name change to CCCA in 1980) by its new masters, the Northern Territory government, but by the Alice Springs regional office of the federal Department of Aboriginal Affairs (DAA). It is by exploring community development practice at this local level, rather than by focusing upon the broader architecture of Australian Aboriginal policy, that the origins of CAT in relation to the dilemmas faced by small Aboriginal communities can best be explained.

Two considerations overshadowed DAA thinking during the 1970s: firstly, how to adjust to the geographical rearrangement of Aboriginal populations in regional Australia, and their altering social needs, as employment opportunities in the pastoral industry collapsed in the wake of the equal pay decision in 1968, the increasing mechanisation of traditional pastoral employment activities, and the broader restructuring of the Australian economy; and secondly, how to respond to the gathering pace of the Aboriginal homelands movement. Whereas the social ill-effects of the equal pay decision upon Aboriginal communities now tend to be glossed over as a natural consequence of marketplace self-correction, the homelands movement has been consistently demonised as an unnatural and unworkable consequence of out-of-touch interventions in traditional Aboriginal society by European 'visionaries' and bureaucrats, which set up Aboriginal outstation communities to fail. Dr H.C. 'Nugget' Coombs in particular has been identified as the architect of this doomed homelands movement.[13]

CAT's origins suggest a different interpretation. DAA's decision to fund an appropriate technologist at ASCC was the result of a convergence of well-informed on-the-ground efforts by health workers, scientists and bureaucrats in partnership with Aboriginal organisations to find practical solutions to Outback settlement needs. This convergence was not without friction: many DAA officials (who were transferred into the new department when it was created out of the old Northern Territory Welfare Branch in 1972) held dated views about Aboriginal society and were sceptical about the homelands movement, most DAA contractors and consultants were out of touch with the culture and social dynamics of the Aboriginal communities in which they worked, and fieldworkers who were sensitive to local viewpoints found themselves 'at the end of a long chain of command and [having] no authority to halt processes which they consider to be damaging'.[14]

Reform efforts were loosely informed by awareness of overseas initiatives in community development and appropriate technology, and the thoughtful adjustment of these concepts to the resourcing and long-term sustainability of Aboriginal settlements in central Australia. However, attention focused mainly upon the accumulating evidence of Aboriginal ill-health, and the slowly dawning realisation that remedial interventions were hampered by 'the apparent mis-match of [European] technology and Aboriginal lifestyle'.[15] It was to address these issues that in 1972 a national workshop, 'Better Health For Aborigines?', was organised at Monash University in Melbourne by the chair of its Department of Social and Preventive Medicine, Basil Hetzel (later famous for his research on iodine deficiency disorders), who in 1970 had also become chairman of the university's Monash Centre for Research in Aboriginal Affairs. The workshop, which was part-funded by DAA, attracted 70 participants, almost a third of whom were Aboriginal representatives. One of them was Yami Lester, a Yankunytjara man from Alice Springs, who was accompanied by fellow Congregational (and later Uniting Church) social justice activist and social worker, the Reverend Jim Downing. They were both members of the Institute for Aboriginal Development (IAD), which Downing had established in 1969 at Alice Springs, and the ideas of both men about appropriate community development and the inappropriateness of existing state-sponsored service delivery and local governance programmes would be important for the early history of CAT.[16]

The Monash workshop dealt with practical issues rather than abstract concepts. It recognised that community development activities must 'take account of different cultural practices within different communities',[17] and emphasised the need for genuine two-way partnerships between Aboriginal communities and outside experts. It drew attention to the poor state of Aboriginal health, emphasised the need for community preventive medicine and community development programmes, recommended 'appropriate' training of health professionals in Aboriginal society and culture along the lines of new courses being developed by IAD, and urged that health planning be undertaken 'in consultation with the Aboriginal communities they are designed to serve ... and carried out through the people themselves and their community leaders'.[18] These recommendations conditioned subsequent thinking on community development in central Australia throughout the decade. Hetzel recalls that the

Monash workshop and its published proceedings 'got the community approach started' and won recognition from Canberra.[19]

In parallel, a series of informal meetings, beginning in 1974, were held in Alice Springs by the Central Australian Aboriginal Congress (CAAC), which had been established by Aboriginal people in 1973 to 'safeguard and promote the interests and rights of central Australian Aborigines', and which embarked upon developing a grassroots issues-based health programme for Aboriginal people in Alice Springs and surrounding homelands communities.[20] One of Hetzel's colleagues at Monash, Trevor Cutter, became foundation director of the CAAC's community health unit. As a result of these CAAC meetings and of the growing momentum of the outstation movement in central Australia, CAAC identified the need for a practical and wide-ranging research and development programme on water, shelter, energy, communication and transport options that were appropriate for outstation communities. Given Hetzel's relationship with Cutter and his high standing after convening the national health workshop in 1972, it was to him that CAAC turned for advice on how to proceed. Hetzel, who early in 1976 had become chief of the new Adelaide-based Division of Human Nutrition in the Commonwealth Scientific and Industrial Research Organisation (CSIRO), was invited to organise a conference in Canberra on 'the ecological problem presented by the longer term needs of these small communities', with an eye to engaging CSIRO in the practicalities of addressing those needs in an ecologically, socially and culturally sensitive manner.[21]

The resulting symposium, held in Canberra in October 1976, was intended to publicise in the national capital the grassroots needs of the homelands movement. Hetzel played a strategic role in the symposium's design and proceedings. On the one hand he applauded the potential of the homelands movement, emphasising that it represented 'decision-making and self-determination' by outstation communities which were seeking 'to preserve elements of their traditional culture and to develop this culture, instead of [succumbing to] the threat of its destruction arising from larger settlements dominated by the white man'. However, on the other hand Hetzel, like Coombs, warned that 'the long term survival of these small communities ... is in considerable doubt'.[22]

A hands-on follow-up workshop attended by Aboriginal community representatives from across central Australia was held in Alice Springs in October 1977. As a result, a coordinating committee was formed, comprising Northern Territory officials and representatives from CSIRO, DAA, the federal Department of Health and the CAAC to foster the development of 'appropriate research projects in central Australia' relating to the practicalities of land management, health and nutrition, and appropriate technology.[23] These initiatives won recognition from the House of Representatives Standing Committee on Aboriginal Affairs, which drew attention to the importance of appropriate technology in its 1979 report on Aboriginal health, and affirmed that the DAA should set up an advisory committee or network to consult across government departments and with Aboriginal communities.

The emerging network became known as the Technical Advisory Group for Aboriginal Homelands (TAGAL). It comprised two groups: a coordinating committee chaired by Hetzel which met periodically in Adelaide, and a subcommittee run by CSIRO and DAA officials in Alice Springs that drew in central Australian researchers and Aboriginal advocates. These included Brian Ede, director of CAAC (and later leader of the Parliamentary Labor Party in the Northern Territory) and Downing from IAD. Downing's contributions were especially important. He had become deeply involved with issues of Aboriginal wellbeing since working as superintendent of the Congregational Metropolitan Mission in inner-Sydney Redfern between 1959 and 1965. Downing's great interest since moving to Alice Springs lay in fostering on-the-ground Aboriginal self-determination in the face of disempowering bureaucratic practices, and to that end he championed both the homelands movement and the participatory application of overseas community development models to practical issues identified by central Australian communities. He advocated cross-cultural training for non-Aboriginal fieldworkers, and the redesign of government policy-making and implementation processes, which he argued were 'just not geared at all to pay heed to feed back from the Aboriginal people'.[24] Downing argued that both of these reforms required the appointment of fieldworkers trained in community development, which he called 'a backward discipline' in Australia.[25]

Whereas Downing emphasised community development, others in the TAGAL network focused upon applied technologies. In 1978 TAGAL in Alice Springs organised an inspection of homelands in the central Australian region in which the Northern Territory, South Australia and Western Australia converged. The report concluded that 'the strongest, Aboriginal-inspired, community initiative we met' was found on outstations rather than at the old mission settlements, but that the homelands movement was being constrained by technical problems with water supply, transport, communications and shelter. Still more worrying, the TAGAL report echoed Downing's concerns by drawing attention to 'a disturbing, two-way lack of knowledge of Aboriginal culture and language by most Europeans and of European ways by most rural Aboriginals'.[26] TAGAL began to map out a detailed research agenda to break down these misunderstandings so as to respond better to homelands needs, from which the emerging priorities – land management, community development, appropriate technology – flowed through into the new position specifications for the Community Management Specialist and the Appropriate Technologist at CCCA in Alice Springs.

* * * * * * * * * * * * * * *

It was ASCC's principal, Jim Pearse, who brought these elements together, combining discussions within TAGAL about applied technology and community development, proposing a centre for appropriate technology to synthesise these approaches, and identifying a funding model to make these innovations possible. In doing so Pearse drew in part on his friendship with Downing, who scaled back his participation in IAD (Lester took over from him as its director) in order to research and write on community development. Pearse provided him with a college office, and benefited in return from Downing's conversations with him about community development and self-determination. Walker remembers Pearse and Downing as being 'in each other's ears'.[27] Pearse increasingly came to the view that the college – its physical arrangements, teaching programmes, student profile (the college comprised 1,500 students, most of them non-Aboriginal and enrolled part time), and the attitude of a majority of its staff – was not conducive to delivering effective technical education to Aboriginal people, and to partnering with them to develop technologies suited to their community development needs. Ede at CAAC was scathing about the college, stating that 'It is common knowledge that people see the Community College as a place for "clever people"', and that there was 'little chance of the College becoming a place where Aboriginal people feel they can drop in and have a yarn about their problems'.[28] Pearse began to think of a counter-campus where these things could happen, and in 1979 convinced the Northern Territory government to buy a run-down site, containing a large shed and outbuildings, in Priest Street, beyond the railway terminus, where an engineering factory, and later a woodworking and cabinet-making factory, and finally a sports store had operated. Pearse proposed that the college could use Priest Street to establish an Aboriginal vocational training facility in pre-trade automotive mechanics, carpentry and joinery, but privately he harboured a more ambitious plan for the site to address the reality that 'the regular college environment was simply not attracting … the Indigenous people of the town'.[29]

Pearse identified appropriate technology as a means of bringing together his sympathy for Aboriginal community development and dissatisfaction with the teaching programmes that the college was delivering. He had read Schumacher's *Small is Beautiful*, and identified Priest Street as a potential alternative campus where not only Aboriginal students could receive technical training, but 'where a centre, for what I then understood, of intermediate technology [could] be developed'.[30] Pearse floated this idea with Dr Jim Eedle, who in 1978 was the founding Director of Education in the Northern Territory, and later Planning Vice-Chancellor at Northern Territory University. Eedle, who had previously worked as an educator in East Africa, told Pearse that 'appropriate technology' was the better term for what he had in mind. Pearse thereupon asked the college librarian 'to undertake a search world-wide for "centres of appropriate technology" to discover, if possible, actual working models. About 1,000 hits were found of which most were of the nature "appropriate technology is a very desirable thing and a philosophy that should be pursued" [but only a] tiny handful of concrete working examples was found'.[31] Pearse nonetheless prepared a discussion paper on the concept and submitted it for the Council of the Darwin Community College to consider. As this was happening,

> the then Commonwealth Minister for Aboriginal Affairs, Ian Viner,[32] visited Alice Springs and through DAA I arranged to meet with him and discuss a proposal, which I had prepared, for the establishment of a centre for appropriate technology. He expressed keenness for the idea ... In discussion the matter of staffing arose and I told him I thought that beginning with two people and going on to six would [be] the way forward. He seemed to go along with this (Remember these were heady days!).[33]

The results of Pearse's exchange with Viner became evident at a meeting in September 1978 organised by DAA in Alice Springs to discuss Aboriginal health issues. Federal and Territory bureaucrats met with Ede and other CAAC members, and representatives from the Pitjantjatjara Health Service, Urapuntja Health Service (Utopia) and the Lyappa Congress (Papunya). Lester and Downing acted as interpreters. Participants identified an adequate water supply as the overriding priority for outstation inhabitants, together with the application of more effective technology to deliver it. They also drew attention to sanitation, food storage and housing. The meeting asserted that the coordination of community services was 'quite inadequate', and 'that information on simpler, more appropriate technology needs to be collected and made available to Aboriginals'. DAA responded that it was providing funds to CCCA to employ an 'appropriate technologist'[34] and an expert in community development. Ede countered that if the positions were based within the college, their activities would inevitably be defined 'in conflict with the people's expressed needs', and he proposed instead 'that the Appropriate Technologist should be incorporated into a resource centre that was under effective community control'.[35] CAAC and TAGAL were both sympathetic to the embryonic Outstation Resource Centre at the Little Sisters Block, on land that had been made available for Aboriginal organisations by the Little Sisters of Mercy on the outskirts of Alice Springs. However, DAA reported to Viner, not Ede, and Pearse had already won the minister's approval for the two college positions.

The position description for Appropriate Technologist that Pearse now prepared noted that the appointee would 'Assist with the development of a Centre for Appropriate Technology at the College', and also stipulated that its 'programs would tend not to be pre-emptive, but rather would arise from consultations with the communities concerned and the assessments made in the light of these by the appointee'.[36] Pearse reiterated these sentiments when, after offering the post to Walker, he spoke to him about his expectations for the new position.

****

Even though Pearse played a key role in preparing the position descriptions and starting the selection process for the two new college lectureships, the changing nature of governance in the Northern Territory influenced the outcome of his initiative. The Commonwealth, which had administered the Northern Territory since 1911, embarked upon a transition period towards Territory self-government during the 1970s. The old Northern Territory Legislative Council was replaced in 1974 by the Legislative Assembly, and elections in October delivered a Country-Liberal Party majority that was out of step with the Aboriginal policies of both the former Whitlam and the then current Fraser federal governments. In July 1978 the Commonwealth's Northern Territory (Self-Government) Act delivered self-government, paving the way for the emergence of a new structure for lawmaking and administration. However, under a Memorandum of Understanding in 1978 between the Australian and Territory governments, the Commonwealth maintained overall responsibility for Aboriginal affairs. In 1979, responding to these developments, John Angel, the regional director of DAA in Alice Springs, instituted regular meetings of Commonwealth and Territory departmental officials to exchange views and coordinate programme delivery. This was the first step in a process by which DAA – which had partnered with ASCC to make the two lectureships possible – gradually stepped back from that relationship and from direct management of Aboriginal affairs in the Northern Territory, devolving responsibility to Aboriginal Community Councils and Territory officials.

Meanwhile, the nature of ASCC was also changing as the college advertised the two new lectureships in 1979. ASCC had grown out of the South Australian Adult Education Centre in Alice Springs. This had initially remained a separate entity when the Commonwealth set up Darwin Community College (DCC) in 1974 and absorbed the staff of the South Australian Adult Education Centre in Darwin into the new college. However, when Cyclone Tracey devastated Darwin in 1975 and destroyed the new DCC campus, its staff relocated to universities and colleges throughout Australia. Many staff moved to ASCC in Alice Springs, which thereafter functioned as a campus of Darwin Community College. Pearse, who had previously lectured in adult education at the University of New England in New South Wales, was appointed head of ASCC in 1977, but was styled a DCC Assistant Principal and reported to its Vice Principal and Principal in Darwin.

With the creation of CCCA in 1979, Pearse became Principal, but in matters of substance rather than name, Northern Territory self-government carried unfortunate implications for him and his priorities for the future direction of the college. Jim Robertson, the Territory's new CLP Minister for Education, was a resident of Alice Springs and took an active interest in its local affairs. It was him and his adviser Geoffrey Chard (a South Australian-trained school teacher working in Darwin) who engineered the separation of ASCC from DCC in order to incorporate it into the Territory's administration rather than Canberra's. Existing college staff were given the option of transferring to DCC or remaining in Alice Springs at CCCA, which became part of a new Technical and Further Education (TAFE) Division headed by Chard within the Northern Territory's Department of Education. Pearse quickly became frustrated that his own vision, for the college to establish a broad community development role in Central Australia with active outreach programmes in Aboriginal education and vocational training, was being derailed by Chard's narrower concept of a conventional TAFE college. The result was that he took leave from the college and never returned.[37]

Pearse was replaced by Howard Tinsley (a former principal of the Goroka Teachers' College, a campus of the University of Papua New Guinea), who had only recently been appointed to the college by Chard. Tinsley was named Acting Principal, an indeterminate position, and understandably echoed the thinking of his TAFE boss in Darwin. His strong-willed deputy, Margaret Crawford, did the same, to an even greater degree. Chard and his subordinates worried that a research centre run by academics with PhDs would be too hands-off to deliver practical teaching outcomes. Rather, they expected that the Appropriate Technologist would complement the college's traditional on-campus teaching programmes – which would include short courses 'identified as being necessary and appropriate in Aboriginal communities' – and new external forms of instruction by 'field-based Adult Educators'.[38] Community educators were appointed at Docker River, Papunya, Yuendumu, Alice Springs and Warrabri, and a regional TAFE coordinator, Tom Marling, was appointed to supervise them. Chard decided that Marling should be based at CCCA, in order to liaise with college staff and coordinate their activities with those of the new community educators (most of whom, in Pearse's opinion, had no appropriate qualifications for their role in Aboriginal communities). This attempted interweaving of the two programmes did not work, and Marling was soon moved to the TAFE regional office in Alice Springs. However, the overall preoccupation with practical TAFE-style training persisted. What the college, DAA, and Aboriginal organisations had earlier agreed upon – a research centre – was now dismissed out of hand.

****

Walker, the young academic from Sydney, thus found himself, even before his appointment was confirmed, in the middle of a dispute that he had not initiated about the nature, location and purpose of his new position. With the college now transferred to the Northern Territory TAFE sector, Chard intervened in the recruitment process for the two new lecturerships and axed the community development position (a specialist at the Papua New Guinea University of Technology at Lae had been short-listed). Chard could not do the same for the lectureship in appropriate technology because Walker had already been offered the position, but Chard phoned him in Sydney to outline his expectations for the Appropriate Technologist. He followed this up with a letter in which he forthrightly laid down his own framework for the role of the Appropriate Technologist, and implicitly invited Walker to resign if he did not agree with it. Chard made it plain that he was 'concerned by your references during our conversation to a "Centre for Appropriate Technology" which I understand has been the subject of discussion between you and the Principal'.[39] Dismissing this concept, Chard sought

> to emphasise, however, that the provision of practical skills training for Aboriginal communities – that is skills which are appropriate to the needs in situ in remote Centralian communities – is what I (and Department of Aboriginal Affairs) believed was the basis for the creation and funding of a lectureship in Appropriate Technology at the Community College of Central

> Australia. I know that the Department of Aboriginal Affairs would have very serious reservations about a program which subordinated skills training to theoretical research, and I would share those reservations.[40]

Notwithstanding Chard's comments, when Walker arrived in Alice Springs in May 1980, he discovered that DAA's thinking was closer to Pearse's than Chard's. Soon after his arrival Walker introduced himself to Commonwealth officials, and 'I mentioned the way Director TAFE saw my role & they disagreed vehemently. They wish me to have as few restrictions as possible. To be administratively responsible to the college admin. but to be able to communicate directly with them and other Govt agencies in the town'. Walker concluded that 'The message was clear that they felt there was a big need for my type of work and that if it did not work at the college I should feel free to approach them and either they would try [to] straighten it out or fund the programme within congress [CAAC] or [the Northern Territory Department of] Community Development'.[41]

At a follow-up meeting in late June 1980 between DAA and college management to formalise the responsibilities of the Appropriate Technologist, DAA officials reiterated 'that the programme be allowed to function as independently as possible of traditional teaching roles at CCCA', and suggested that 'for the sake of identity and improved communication with aboriginal groups and outside contacts it would be useful to have a title – like Centre for Appropriate Technology at CCCA – even though it might exist in name only'.[42] However, when Walker tabled a paper that advocated establishing just such a centre, differences of opinion immediately surfaced. Marling, TAFE's regional coordinator, objected and, obliquely distancing current Territory thinking from Pearse's earlier advocacy of a centre, 'stated that Dr. Walker's paper and other literature may incorrectly suggest that a Centre for Appropriate Technology would be set up at the C.C.C.A. but that it should be understood by all concerned that such a centre would not be established in the foreseeable future'. Crawford, CCCA's Deputy Principal, was more transparent, stating bluntly 'that the Director T.A.F.E. [Chard] had given instructions that a Centre for Appropriate Technology would not be set up'.[43]

CCCA displeasure was reflected in Walker's relocation to the periphery of the college's Anzac Hill campus, where he was allocated a demountable by the Todd riverbed. Walker's marginalisation from college activities was taken a step further when Tinsley advised staff in July 1980 that 'Bruce Walker will ... be working closely with Tom [Marling] in the planning and implementation of Appropriate Technology for the region', rather than with college senior management.[44] However, in early September Tinsley went a step too far when he sought to have Walker relocated off campus altogether, to Marling's TAFE office in Alice Springs. Tinsley advised Walker that he would henceforth report directly to Marling, and that although the college would continue to support him, 'these changes firmly site you within TAFE and therefore logically your physical location should follow suit'.[45] Walker formally protested against this 'ultimatum presented to me', pointing out that he had been appointed as a staff member to the college and not to TAFE, and that his ability to continue to use college facilities in order to develop an effective appropriate technology programme would be 'based on goodwill alone'.[46]

Chard intervened, perhaps mindful that Walker's position was funded by DAA officials who were open-minded about where the position might be best located, and that any unilateral variation of the Commonwealth funding agreement would probably be illegal. He instructed Tinsley that

> Dr Walker is a member of the College staff who is concerned with the provision of a service specifically to Aboriginal communities in Alice Springs and, especially, in outlying areas. As a staff member of the College he is both professionally and administratively responsible to the Principal but, naturally, I would expect a very close liaison to be worked out with the Regional Co-ordinator [of TAFE] who, through his network of Adult Educators based on-site in communities, should be the focal point for the identification of needs to which Dr Walker should respond.[47]

Chard offered a pragmatic way forward, which paradoxically put back on track Pearse's agenda for a college counter-campus that Chard had hitherto done so much to derail. Chard suggested that Walker be moved off campus to the college's near-moribund Priest Street facility, given that his role was 'the provision of practical training in skills which are appropriate to the needs in situ in remote Centralian communities', and therefore that 'he would be best located with staff engaged in a vocational training program from whom he could draw support and offer the benefit of his experience and expertise'.[48]

CCCA senior management digested this setback to their plans. The Appropriate Technologist was at least to be moved off campus, and the Priest Street property was nothing more than a large derelict shed. They announced that Walker could have a small partitioned space at the front of the building and that the college's lecturers in automotive mechanics would occupy the rest. Tinsley continued to attempt to impose a traditional TAFE model upon Walker's activities. He wrote to Walker in November 1980 that he must put more energy into developing training courses, rather than visiting remote communities and taking note of their technology needs, and he cautioned Walker in March 1981 that 'whereas I acknowledge the importance of the resource aspect of your brief I would be reluctant to see this overshadow the practical skills training component'.[49] Counteracting the advice given to Walker by DAA, Tinsley wrote that 'we have discussed this previously, when I conceded the need for familiarization activities, evaluation of needs and the establishment of appropriate contacts during your first months. I would hope to see this investment bear fruit in a series of specific courses within this year's programme'.[50] Walker retorted that he was designing the new Appropriate Technology programme 'to meet needs recognised in communities' rather than imagined by college bureaucrats, and that this 'is what distinguishes the A.T. programme from aboriginal training programmes offered in the past'.[51]

Tinsley resigned as Acting Principal late in 1981, and Bob Cruise was appointed Principal early in 1982. He was much more sympathetic to Walker, and in staff circulars began to refer to Walker as head of the 'Centre for Appropriate Technology'.[52] Cruise was not contradicted by Chard, whose attention was focused elsewhere, and who in 1983 moved to the newly established Vocational Training Commission. Cruise ruled that Walker should be in charge of the entire Priest Street property and its staff (including the lecturers in automotive mechanics). Walker moved from Anzac Hill to Priest Street in June 1982. One of the first things he did was to 'put up a piece of tin on the front that said "Centre for Appropriate Technology"'.[53]

2. Early Development

CAT began with just one staff member and a hand-made sign hanging off a derelict shed in Alice Spring's industrial area, far from CCCA's main college campus. However, in dovetailing aspiration with achievement, CAT's early development benefited hugely from the fact that college management effectively allowed Walker a free hand in establishing a wide-ranging appropriate technology programme, because they never visited Priest Street to check what was happening there.

CAT enrolled its first students in September 1982, teaching the college's pre-existing Aboriginal pre-apprenticeship automotive and mechanical skills course. Walker soon added flexible one-week courses in welding, small engine repair and preventive maintenance for motor vehicles. However, Walker's notes from this time show that he was already alert to the 'Difficulty of Training Bush people in Town', and to the imbalance between the few relevant college courses on offer and the 'Wide range of Aboriginal needs'. He was also becoming aware that among college staff, 'everybody reckons they have the

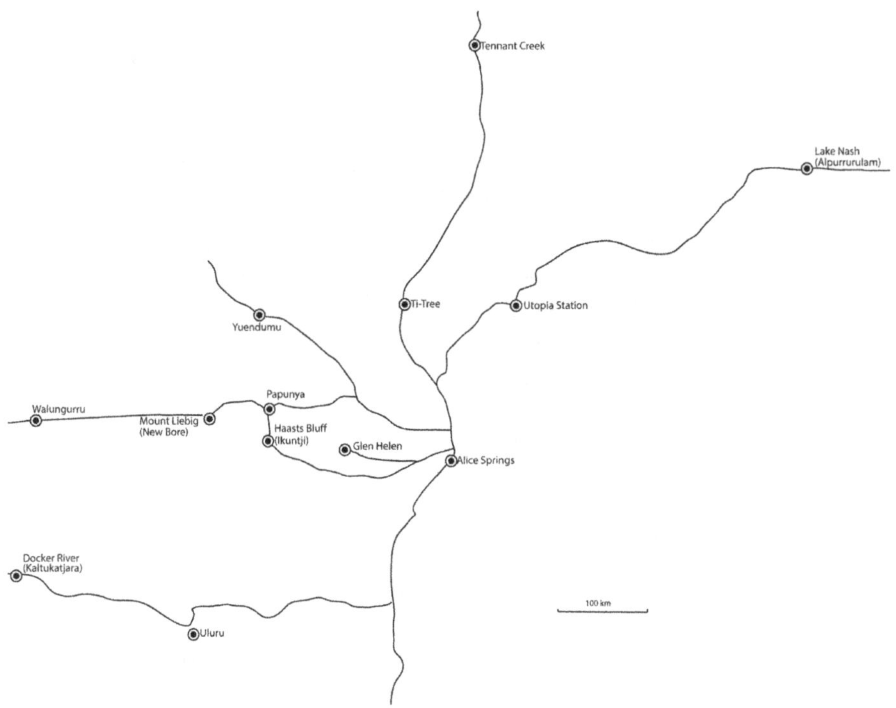

Map of the central Australia region covered by CAT's early work.

solution to Aboriginal problems' and that they were therefore 'often inflexible in [their] approach to training'.[54] In addition to providing technical training at Priest Street, Walker set up his own workshop and a display area where visitors could see the water and energy management technologies with which he was experimenting. However, he hesitated to develop more comprehensive training courses and alternative technology displays until he could learn from Aboriginal communities in central Australia what they really wanted.

From the moment he arrived in Alice Springs, Walker was encouraged by the Commonwealth and Territory officials who worked directly in the homelands to engage with Aboriginal communities. He was given contact names in Utopia, Papunya, the Tennant Creek Housing group, the Pitjantjatjara Council, Finke River Mission, the Elliott-Newcastle Waters Housing group and Yuendumu. Staffers in DAA and the Northern Territory Department of Community Development invited Walker to join them on their field trips, and urged that in carving out a role for CAT the 'Immediate impact [of appropriate technology should be felt] in homelands settlements' rather than simply put on display at Priest Street.[55] Walker agreed, and in

The TAGAL group inspecting an early handpump at Blackwater near Papunya.

May 1980, within weeks of arriving in Alice Springs, he joined Department of Health staff on a trip to Papunya, about 250 kilometres west of Alice Springs, and to the surrounding outstations in the Haasts Bluff (Ikuntji) region of the West MacDonnell Ranges. In the following month he accompanied DAA staff on a long excursion north-east of Alice Springs to the Utopia station, which had been purchased for the traditional landholders in 1976. He would travel regularly in the wide hinterland around Alice Springs and beyond over the months and years to come.

Walker simultaneously engaged with existing research networks in central Australia. In September 1980, when DAA organised another TAGAL workshop – which was held in a dry creek bed at Blackwater, a new outstation near Papunya – Walker 'tagged' along to it.[56] There he met local CSIRO scientists and experts in other government agencies, and as a result joined the Alice Springs Working Group of the Interdepartmental Water Needs Committee.

When the federal government announced additional funding through DAA for water supply, sanitation and power in Aboriginal communities, starting from mid 1981, Walker participated in the inter-departmental committee that recommended funding allocations. He also built up relationships with university research networks, maintaining close contacts with his alma mater, the University of New South Wales, and established new ones with the Australian National University's Northern Australia Research Unit in

Darwin, the Faculty of Engineering at the University of Melbourne, and Murdoch University's Institute for Environmental Science. As a result of Walker's hard work, when the next big TAGAL conference was held in Alice Springs in October 1985, Walker (supported by DAA, CSIRO and the Central Land Council) was its chief organiser. Walker also tapped into international research and community development networks, contacting other experts overseas, building up a library of technical reports, and undertaking fieldwork for AusAid in Tanzania in 1983, Burma in 1985 and 1986, Kenya in 1987 and Vanuatu in 1987.[57]

****

During Walker's first two years in Alice Springs, from his arrival in May 1980 to his establishment of CAT at Priest Street in June 1982, he had followed Jim Downing's advice about what any serious community development worker must first do: 'sit and learn from [the] community for six to twelve months with no pressure to produce anything'.[58] CCCA senior management sought to impose a traditional classroom approach, but the results of Walker's apprenticeship became quickly evident after his move to Priest Street, and the independence from college oversight that this made possible. These results can be seen in five key areas of early CAT activity.

Firstly, CAT energetically engaged in applied research that could deliver tangible results to Aboriginal communities. TAGAL's 1980 workshop concentrated upon the practical issues of providing effective water management, energy, shelter and communications technologies for remote settlements, and Walker targeted these areas in his emerging research programme. TAGAL's workshop had agreed to undertake a detailed land management case study in the Haasts Bluff area, combining staff expertise and resources from some 20 agencies (including CSIRO, the Department of Primary Industry, IAD and the Central Land Council) and Aboriginal community organisations. TAGAL's objective was to create a detailed knowledge base and work programme 'for the development of an ideal "outstation" in the Haasts Bluff / Papunya area'.[59] Walker became an active participant in this initiative. He simultaneously became involved with the Lake Nash (Alpurrurulam) outstation movement, which Downing had championed since the late 1970s. DAA requested CCCA in January 1981 to allow Walker to assess water purification options for drinking water at the Lake Nash outstations, on the Queensland border north-east of Alice Springs. Lake Nash had been an embarrassment for the federal and Territory governments since 1978, when the pastoral company there sought to relocate the Aboriginal community that had previously worked for the cattle station before the equal pay ruling to 'poison gidgee bush country'[60] on the far southern boundary of their lease. Trial bores found high nitrate levels in the local water supply. For the next 20 years, Walker and his colleagues would exhaustively study the quality of water supplies available to remote Aboriginal communities.

Secondly, CAT began to design and manufacture appropriate technology products for outstation use. The trigger for this development was Walker's participation in TAGAL's fieldwork in the Haasts Bluff region, together with his early visits to Papunya, the welfare settlement to which Pintupi people from the Haasts Bluff mission had been moved in 1960. Walker's introduction to this region coincided with the beginning of the outstation movement by the Pintupi out of Papunya to their Western Desert homelands. However, resettlement was hampered by the difficulties of supplying water to distant sites, and outstation groups had to truck water from Papunya in 44-gallon drums. The outstation coordinator, Allen Jenkins, met Walker in Papunya and asked if it was possible to tap into some of the government bore heads that were being established alongside roadways in the region. Walker undertook preliminary design work at the CCCA workshop, and then visited the New Bore outstation, to the west of Papunya near Mount Liebig in the MacDonnell Ranges, in late July and early August 1980, during which time he rigged up a prototype hand pump out of star pickets, with help from Jenkins' team of outstation workers. After several days of cutting steel, welding, and fine-tuning, Walker recorded proudly on Saturday 2 August, 'Took photographs of completed pump and extension arm'.[61]

Within a week, Pintupi re-settlers had occupied the bore site. During August 1980 Walker improved the design and custom-built a new hand pump at the CAAA workshop, returning in October to install it. He found some 30 people living nearby. The Papunya outstations requested more hand pumps, plus training in their installation. By early 1981, when Walker installed a hand pump at the new Walungurru (Kintore) outstation, near the Western Australian border some 520 kilometres west of Alice Springs, eight hand pumps were already operating in Papunya outstations that stretched westwards as far as Kiwirkurra in the desert regions towards Port Hedland. At Kiwirkurra, Walungurru, Ikuntji and neighbouring outstations Walker thus provided the initial basic settlement infrastructure that underpinned the famous Papunya or Western Desert art movement. The Pintubi artists who had begun to work with Geoffrey Bardon in the early 1970s to kick-start this movement were the same men who collaborated with Walker to install hand pumps in and around Walungurru during the early 1980s.[62]

Alan Jenkins, Papunya Outstations Resource Centre Co-ordinator, testing the original star picket prototype handpump at Mt Leibig in the winter of 1980.

After Walker's relocation to the big workshop at Priest Street in 1982, CAT was able to support requests from outstation communities for an increasing range of products that were delivered across a widening range of locations that included Walungurru, Docker River, Yuendumu, Utopia and Lake Nash. The most pressing initial needs – often identified by women – were water and energy supply, shelter and waste management. In responding to homelands demand, CAT had by the mid 1980s expanded from Walker on his own to 22 staff and was working with 40 Aboriginal communities in the Northern Territory, Western Australia and South Australia. By the end of the decade the list had grown to some 70 communities. In addition to installing hand pumps, CAT developed wheelchairs, hand-powered washing machines, woodchip water heaters (developed at Utopia), outside ovens, an outdoor shower and laundry unit, and, 'its *piece de resistance*[,] the ventilated improved pit latrine – proudly dubbed VIP for short' (developed at Docker River).[63] These have all become icons of Outback Australia. CAT produced over 700 chip heaters and almost 1,000 VIP latrines for remote locations in the Northern Territory, Western Australia and South Australia during the 1980s.

Thirdly, CAT sought to provide practical advice about community use of alternative technologies. Thus, in tandem with his development of appropriate technology products, Walker distributed technical notes to describe their

Nugget Blackmore was appointed as janitor at Priest St and assisted in converting the run-down sheds into the CAT enterprise training workshop.

installation and use. In February 1981, for example, Walker wrote to DAA in Darwin that he had 'almost finished the production of a little booklet on the handpump' that he had designed for outstation use in 1980.[64] A booklet on the hand-powered washing machine was produced in 1983. CAT produced its first comprehensive handbook, *Appropriate Technology in Central Australia*, in August 1982, which provided details about water, energy, shelter and communications options for remote communities. An enlarged second issue was released in August 1984. Walker collaborated with CSIRO scientist Barney Foran to publish TAGAL's 1985 conference papers as *Science and Technology for Aboriginal Development*. Walker also gave public demonstrations. In June 1985, for example, he travelled to Ernabella in South Australia to act as an instructor for a community workshop organised by the Anangu Teacher Education Programme and, in 1986, CAT's exhibits at the Darwin Show attracted considerable interest. Influential visitors began to tour Priest Street, including the High Commissioner for India in 1985 and the House of Representatives Standing Committee on Aboriginal Affairs in 1986. The Australian Broadcasting Corporation's national television programme *Quantum* carried a story on CAT and its work at Walungurru (Kintore) in 1988, triggering a further round of inquiries and orders from the CAT workshop.

The early years at Priest St making the most of the existing structures.

Fourthly, Walker, assisted by Nuggett Blackmore, the Aboriginal janitor appointed to Priest Street, transformed the CAT workshop from a one-man show into a major production facility with a significant technical training function. In February 1983 Noel Bowen, whom Walker had met during 1980 when Bowen worked for the Aboriginal Tangentyere Council in Alice Springs,

The emergence of CAT as a National Technical Resource Clearinghouse necessitated the development of more appropriate office and work spaces. The monolite wall was built by men from the Kintore building team.

began volunteer work as a technician at the new Priest Street workshop to help Walker meet the growing demand for appropriate technology products. DAA intervened to fully fund Bowen's position as a Technical Officer from September 1983. The Department had been embarrassed to find that Bowen, unpaid, was working full-time as manager of a booming production workshop that had also since May of that year been successfully training Aboriginal technical workers.

The workshop's transformation from a small-scale experimental site into a full-scale production and training facility had been assisted by the Fraser government's announcement in January 1983 that as a result of a funding agreement between the federal, state and territory governments, a Wage Pause programme would be launched to support training schemes for unemployed youth and the older unemployed. CCCA Principal Bob Cruise drew Walker's attention to the scheme shortly before applications were due, and together they prepared a rushed proposal to provide Aboriginal technical training in the CAT workshop, simultaneously meet the growing demand for CAT products from the Aboriginal homelands, and educate Aboriginal technical workers with the basic skills to operate small community workshops. Their application was successful and a $50,000 grant was awarded, enabling CAT in May 1983 to begin employing three adult and three junior Aboriginal trainees (including two women) under Bowen's supervision to produce hand pumps, hand-operated washing machines, solid fuel water heaters and mobility aids. Among the first trainees to be employed were Herbert Bloomfield and John Parfitt, who would become long-term employees at the Priest Street workshop. CAT's Wage Pause programme generated almost $10,000 in orders within the first two months of operations.

Wage Pause, however, was by then already an obsolete programme, because in March 1983 the Fraser government had been defeated in the federal elections and replaced by Bob Hawke's Labor government. The incoming government shifted the policy goalposts, unveiling a new job creation strategy through its Community Employment Program (CEP) that started in July 1983. Wage Pause funding was rolled over into the new scheme, and in early 1984 CAT was awarded a further 12-month award of $80,000 through CEP to gear up a fully-fledged 'Appropriate Technology [Enterprise] Training Workshop' at Priest Street 'where Aboriginal people may be trained and involved in the production of articles for their communities'.[65]

Priest Street's CEP-funded workshop quickly exceeded expectations, generating $152,000 in orders during its first three months of CEP-funded operations. By the end of the year the workshop had produced 40 VIP latrines, 55 chip heaters, 12 washing machines and 10 hand pumps, and a diverse range of other outstation products for use in the Northern Territory, South Australia and Western Australia. The workshop's production was not speculative; it was sustained by orders received. As Walker remarked in 1984, 'he was having more and more people coming into his workshop getting items off their own initiative'.[66] Moreover, community representatives began to visit the workshop in order to work alongside and learn from the Aboriginal trainees, and an informal campsite – first occupied early in 1984 by visitors from Kintore – soon developed at Priest Street, and this family presence encouraged the trainees to persist with their apprenticeships. By the time CEP funding ceased in March 1985, CAT's workshop had become financially self-supporting, and the number of trainees employed there had increased to ten. Towards the end of the decade it was called 'probably one of the most amazing workshops in the country'.[67]

Dorrie Wesley and Trevor Corbett being interviewed by Anne Deveson during the mid-1980s.

Priest Street in these early years was a rough and ready place. Walker issued an internal circular in 1985 complaining that working conditions were 'deplorable [with] tools and equipment ... just dropped in the dirt' and needing to be regularly replaced as they became damaged or disappeared.[68] He confided to a friend in the following year, 'Tony is still drunk. Charlie has admitted he did the wrong thing giving the generator away. Alec has resigned and Martin has gone missing ... Just like a regular little TV mini series'.[69] But amid such tensions and misadventures, an unambiguously successful Aboriginal production and technical training facility took root. Its dynamic workplace culture was shaped as a diverse collection of people worked together on novel and practical tasks, producing not only tools and implements but a new 'real time cross cultural relationship'.[70] This distinctive culture was perpetuated in jokes and stories that conditioned the thinking of new staff as they joined CAT during the 1980s and 1990s. Aboriginal people who were employed in the workshop still reflect fondly on the experience. Tony Renehan, then a young welder, says 'we were drunks, we were somewhat unreliable but we did good work and look at us now!' Dorrie Wesley and Trevor Corbett, interviewed by Anne Deveson during the mid 1980s, recalled that it was the first time they had made things that other people valued, paid good cash for and then used to good effect. At that time there were limited opportunities for Aboriginal people to 'make something that could be sold' and thus to experience what it was to be 'productive' in mainstream Australian terms.[71]

Fifthly, CAT added on-site training to its workshop production and outstation installation activities. As the volume of products grew from the Appropriate Technology Enterprise Training Workshop, Walker successfully requested during 1984 that CEP funding also be used to begin an Appropriate Technology Extension Service in the homelands, because as he explained, it 'is physically impossible for us to carry out this work without commencing an extension service to work with and train people in communities where they are installing the equipment they have bought from the workshop'.[72] Walker was thus able for the first time to expand CCCA technical education programmes successfully beyond Alice Springs to the Aboriginal homelands. As Walker explained to the Docker River Council in 1984, 'Unlike contractors, our role is to work with your Community so that you know how the showers were built and how to fix the simple things that can go wrong'.[73]

By 1985 CAT's extension service had become self-funding, and together with the workshop generated a substantial surplus. This made possible the purchase of rugged vehicles for installation work and the appointment of additional staff. The main activities and core people at Priest Street were thus funded from sources other than the college to which CAT belonged. This anomaly enabled CAT to build more readily upon its early experiences with the workshop in order to formulate an ambitious Aboriginal Technical Worker training programme (ATWORK), which sought to broaden the TAFE sector's specialist apprenticeship training approach to provide across-the-board practical skills-acquisition training that assisted both personal development and the self-reliance of homeland communities. CAT argued that the existing trade-based TAFE training system effectively excluded 80–90 per cent of potential Aboriginal applicants.[74] In 1986 the national TAFE sector awarded CAT some $360,000 to test its provocative proposal through a two-year feasibility study involving 34 Aboriginal communities. A recent graduate from Walker's own

CAT extension services supplied 14 ATAF ablution blocks and 14 VIP latrines (one at each house) for the equivalent of the replacement cost of a single community ablution facility that was not functioning and did not meet the Docker River residents' desire to have ablutions at each house.

undergraduate degree programme at the University of New South Wales, Kurt Seemann, whom Walker interviewed on the platform of Sydney's Central railway station, was employed as a research officer at CAT in January 1987 to develop this project. As Seemann remembers events, no sooner had he arrived than

> Bruce said, OK I am off to do AusAID work in Africa for several weeks. Can you set up a national group of Aboriginal Educators, and organise a national steering group workshop please? Then he left. No other details. In fact, not even the CAT staff knew what I was here for and thought at one stage I was planted by Bruce to watch them. So with the benefit of not knowing what I was not supposed to do, I boldly [proceeded].[75]

Seemann's subsequent feasibility study would effec tively launch CAT's most important education and training activities.

At an ATWORK national seminar in 1987, held to help plan CAT's feasibility study, Walker began proceedings by announcing bluntly that since its beginnings in 1980, CAT 'has spent the last seven … years addressing the apparent mismatch of technology and lifestyle in remote Aboriginal communities and attempting to design pieces of hardware to facilitate greater Aboriginal involvement in community lifestyle' and that in so doing, they had realised that if technology was well designed and appropriate to its setting, 'there are many new opportunities for work and training which hitherto have not existed'. ATWORK was intended to seize that opportunity. The ensuing workshop emphasised that in order to address the training black hole that CAT had identified, the ATWORK feasibility study must follow 'a meaningful and consultative dialogue with the recipients of the proposal'. The workshop cautioned that 'Community self-sufficiency and the maintenance of a sense of Aboriginality or "spirit" underlying the values of the technologies used (both traditional and introduced) ought not be overlooked'.[76]

The beginnings of ATWORK, together with the ongoing operations of the workshop and the extension service, made Priest Street a busy place, and its backyard campsite a near-permanent place of residence, full of 'Aboriginal people in swags'.[77] When the federal Labor Minister for Education, Senator Susan Ryan, visited CAT she expressed shock that students and their families should live in such conditions, and provided funding in 1988 for a student residential complex for up to 24 people to be built. Jim Bray was appointed as residential manager.

The CAT Residential Complex in Priest St provided in-town support for students and visitors from outlying communities in the region.

Bray was an Eastern Arrernte man from Alice Springs, then aged in his late 40s. He had had a varied career, labouring in timber mills in South Australia as a young man, fruit picking along the River Murray and training as a welder. Bray took on bush jobs and railway work in the Northern Territory and became a skilled arc welder on drilling rigs, working as far north as Nhulunbuy. He also managed Aboriginal hostels in Katherine, Darwin and Alice Springs. During the mid 1980s, when working at IAD, he brought groups of students from its community development programme to see what was going on at CAT. Walker admired him, and thought of him when he needed a manager for the new residential complex. Bray recalls with a chuckle that he 'got interviewed and got a job while shopping at the supermarket': 'I was living over the east side with family, and I went down to the shopping centre to get some stuff, and ran into Bruce there, and he was down a couple of aisles away from me, and … the first part of the interview was really conducted from several of the food aisles. He was down [aisle] three and I was in five or six … and Bruce more or less said … come down and have the interview and we've got a job there for you'.[78]

It was a hard job, because the residential complex was immediately heavily used. Bray recalls that the job was 'seven days, 24 hours', with people arriving at the gate 'charged up and aggro', and 'they'd want to fight you (chuckles)'. Bray 'didn't get a lot of sleep when we had a full house'. He enforced a 'no grog, no drunks' policy, but also took the students under his wing, helping them to cook and preparing them curries.[79] The residential complex quickly became a showcase, and Bray won wide respect and affection.

3. Consolidation

CAT, a concept that TAFE officials rejected out of hand in 1980, had become the strongest organisational unit within Central Australia's TAFE sector by the mid 1980s. One man's determination in 1980 was shared at the end of the decade by a staff of between 15 and 23 people (half of them Aboriginal). From an operating budget of almost nothing in 1980, CAT's annual operating budget by the end of the decade was nudging $1 million (in 1987 it was made up of approximately $100,000 from DAA, $120,000 from the Northern Territory Department of Education and $600,000 generated directly by CAT).[80] CAT's 'hard grunt "doing work"' during these years established the community development interrelationships that enabled the organisation to expand rapidly during the following two decades.[81]

However, success also caused growth pains. Mounting profits enabled CAT to staff and resource its activities better and with greater independence from CCCA, but CAT's activities did not correspond with the College's expectations for the running of not-for-profit educational institutions. As Walker conceded in 1984, 'Whilst we initially sold training produce [without provoking adverse comment] it is hard to justify $½ million worth'.[82] As CAT's revenue increased and its activities expanded, so the accounting and personnel restrictions imposed upon it by TAFE regulations became more burdensome. College red tape hampered research development, and it especially compromised CAT's community development activities. As the minutes of one college council meeting recorded, CAT 'is employing Aboriginals on the settlements. These employees expect to get their money in their hand as soon as the job is completed. If they do not get this then they will not do the job. There is no vehicle in Government Accounting whereby you can pay people on the spot. This problem cannot be overcome in a normal departmental structure'.[83] The different mindsets of college bureaucrat and outstation resident seemed impossible to reconcile. Cruise and Walker organised a series of meetings during 1984 to consider options for CAT's future. These included making CAT an incorporated benevolent association or an incorporated company, but eventually it was decided to leave things as they were because achieving fundamental institutional change seemed 'too hard'.[84]

Three developments changed this thinking. Firstly, Cruise stepped down as principal in 1987, and in the same year the college was restructured as Alice Springs College of TAFE (ASCOT). This closer realignment with the mainstream TAFE sector in the Northern Territory took CAT's relationship with the college back to the uneasy days when Tinsley was acting principal in the early 1980s. CAT's culturally nuanced research and development and off-campus training programmes jarred against what Downing in 1988 called 'the unrealistic demands of an Education Department that seems culturally unaware'.[85] The differences in approach between the college and CAT came to head at the council meetings of ASCOT. On the one side Walker used these meetings to complain about insufficient support from the college and the Department of Education, and asserted that 'the political will for CAT to succeed is not here. Much is said about CAT but little is actually done when it comes to the crunch or positive commitment'.[86] On the other side some council members complained that CAT acted without sufficient reference to the college, which posed the danger that ASCOT might fragment into 'distinct units rather than one cohesive College'.[87]

Secondly, in 1986 the Federated Miscellaneous Workers Union (FMWU) served a log of claims against the Northern Territory Department of Education, claiming that CAT's Aboriginal trainees should be considered as public servants rather than private employees. The Department of Education successfully defended its position, in a protracted case that dragged on into 1987, but behind the scenes the Departmental Secretary recommended that further union challenges could be prevented by accepting Walker's earlier proposal that CAT be made an incorporated association, with a separate board of management that nonetheless remained affiliated with ASCOT. The Minister for Education agreed, with the proviso that 'the Department retain control'.[88]

Thirdly, by the late 1980s it was becoming clear that at a federal level DAA would be restructured and that the new agency that replaced it was likely to fund only Aboriginal organisations. Walker put it to the Department of Education in Darwin and ASCOT in Alice Springs that DAA funding for CAT would be lost unless the latter was restructured into an incorporated body with an Aboriginal board of management. ASCOT accepted this proposal in 1987, albeit with CAT remaining affiliated to the college and an ASCOT representative sitting on the CAT board. In November 1989 CAT was formally incorporated under the Northern Territory *Associations Incorporation Act 1978*.[89]

* * * * * * * * * * * * * * *

CAT's widening operations encouraged Walker to think beyond issues of college governance and begin to engage with broader governmental policy on Aboriginal affairs, as he saw first hand in the Northern Territory the mixed results of the Whitlam government's reformist emphasis on Aboriginal self-determination, and the Fraser government's more cautious emphasis on Aboriginal self-management. Notwithstanding good intentions in Canberra, it seemed that in Outback communities Aboriginal people were becoming locked into new patterns of dependency.

Walker struck a gentleman's agreement with Lester at IAD that CAT would not duplicate IAD's community development teaching, but would complement this by focusing upon the technical aspects of homeland settlement. However, Walker and Ede at CAAC both expressed disappointment that DAA's implementation report following TAGAL's 1980 workshop pledged no additional homelands funding for such activities, and Walker proposed that he and Ede join forces at the local level to lobby for a better deal, predicting that it 'seems that any advance on what we have already will have to be pushed from this end'.[90] Walker also complained that TAGAL's fieldwork at Haasts Bluff was being held back because of inadequate funding for research and take-up by Aboriginal organisations. He contended at TAGAL's 1985 conference (to which he astutely invited the influential homelands advocate H.C. Coombs) that TAGAL's Haasts Bluff-Papunya fieldwork had nonetheless demonstrated that applied research 'can make a contribution to the quality of life in Aboriginal communities, but only if it is supported beyond the research stage'.[91] Walker likewise used the findings from his early water analysis at Lake Nash to urge the Interdepartmental Water Needs Committee 'that a number of communities are presently living on water supplies that are considered unacceptable by the Department of Health', and he succeeded in making governments acknowledge that high nitrate concentrations were a major problem in Central Australia.[92] In July 1989 Walker's technical expertise and policy knowledge were recognised when CAT won a major research consultancy from the Human Rights and Equal Opportunity Commission to assess the provision of water to Aboriginal communities throughout Australia.

However, the force of Walker's arguments, and indeed the continuing success of his early initiatives at CAT, was constrained by the vagaries of altering government policy and funding mechanisms. CAT's workshop and extension programmes were dependent upon the Fraser government's Wage Pause scheme, and then the Hawke government's CEP funding formulas. Although CEP funding in particular consolidated CAT's early growth, CEP red tape controlling the employment of trainees at the Priest Street workshop and payment for community participation in the installation of workshop products became irksome impediments to CAT's expanding activities. Moreover, by the time CAT's activities became self-supporting in the mid 1980s, federal funding for the Northern Territory had begun to decline, and although expenditure on Aboriginal welfare continued to increase throughout the decade, Walker had to forge additional partnerships with service providers and funding sources within the Northern Territory bureaucracy. Yet as one Northern Territory politician complained in the Legislative Assembly after visiting CAT in 1984, 'very little support is being given to [CAT] by the Northern Territory government'.[93] In 1989 DAA, the federal agency that had funded Walker's position in the first place, was abolished as a new federal government policy environment began to take shape at the end of the decade. CAT would need to find new friends in Canberra as well as Darwin in the coming decade.

Federal funding during the 1980s had underwritten CAT's initial development. On the supply side, federal dollars enabled the establishment of CAT's workshop and technical training activities. On the demand side, federal funding also gave Aboriginal communities the ability to buy CAT products and pay for technical education services. During the 1980s social security payments and other government assistance programmes became readily available to all Aboriginal Australians regardless of their place of residence. However, in important respects this empowerment of Aboriginal communities was more apparent than real. CAT's development during the middle and later years of the

1980s, which challenged people to rethink their assumptions about technology, culture and their interrelationships, offers a different and largely overlooked perspective on the paradox during the decade of increasing signs of Aboriginal empowerment on the one hand, and growing expressions of Aboriginal frustration and alienation on the other.

****

Walker mused in 1985, 'It is clear to me that it is the human dimension of science and technology about which the least is known'.[94] The homelands movement and subsequent government attempts to provide remote settlements with community infrastructure highlighted the disjunction between mainstream Western technologies and their cultural reference points, on the one hand, and the needs and aspirations of remote Aboriginal communities on the other. As CAT grew and learned from its experiences in Central Australia, it consistently drew attention to 'the inability of people who live in remote communities to access science and technology in an easy and meaningful way … [and] the inability of many scientists and technologists to … relate to people's needs'.[95] CAT emphasised that these outcomes were the result of cross-cultural misunderstanding, not of cultural problems specific to Aboriginal communities.

Unmodified European urban technologies (for example, flush toilets, electric washing machines, solar panels and house designs) do not work well in small, remote communities. Sometimes the incompatibilities are clear cut: washing machines designed for periodic use in single-family suburban homes break down after continuous use by extended families with heavy loads of dirt-filled ground blankets. Sometimes the incompatibilities are more subtle. Criticising the settlement and housing designs that gradually prevailed at Walungurru over a decade of experimentation, CAT argued in 1990 that the 'guiding principle in Kintore's housing appears to have been the Australian dream of home ownership transplanted to the desert. This model is based on a nuclear family living on a quarter acre block, in a street of like-minded citizens. This model limits the ability of the people of Kintore to practice their chosen lifestyle'.[96]

Many well-intentioned science and technology interventions in remote settlements during the 1970s and 1980s had the unintended effect of locking in Aboriginal dependence and locking out their participation because the introduced technologies were unmediated by local advice about the situations to which they were applied. At TAGAL's 1980 conference, Walker noted that 'the predominant feeling of the scientists and technologists involved was that technical solutions already existed to almost all basic problems faced by

Aboriginal people … [and] that the only additional requirement was to make the solutions work in an Aboriginal context. This also had an easy answer – "we need to educate *them* to make it (technology) work"'. Walker cautioned that this patronising thinking 'reduced Aboriginal participation with the technology. It has locked them out of adapting the technology in a[n] innovative way and transferred their energy expenditure from a dynamic and active lifestyle to a comparatively static and sedentary lifestyle'.[97]

Walker called existing state-sponsored interventions to impose such design principles and products 'a form of technological imperialism' that was undertaken without regard for the diverse environmental, social and cultural contexts of the settlements in which they were placed.[98] His critique was not based upon Eurocentric assumptions that the take-up of sophisticated imported technologies was stymied by a static and technologically primitive culture, but upon his recognition of legitimately alternative and environmentally appropriate Aboriginal lifestyle practices and emphases that went unheeded by policy makers and technologists. Shortly after arriving in Alice Springs in 1980 he declared that

> Appropriate Technology recognises that different cultural and geographical groups have different technologies that are appropriate to their circumstances; that technological self-determination is essential to cultural identity. It suspects that the only wise technologies are those which seek to accommodate themselves to the natural, physical, psychological, economic and cultural environments within which they are used.[99]

Towards the end of the year, having travelled extensively through the homelands communities of Central Australia, Walker argued that

> One of the aims of CAT is to identify local traditions and attempt to use them as a basis for development of new but somewhat related technologies. The idea is not to preserve the past at all costs and turn remote settlements into museums, but to help those who have generations of experience behind them make the transition and adapt in an innovative way to the new needs of the community.[100]

Walker thus endorsed what TAGAL fieldworkers and others who worked directly with homelands communities were already saying, that the outstation movement did not – as media commentators, politicians and bureaucrats in the distant capital cities increasingly said it did – represent a return to traditional practices frozen in the distant past and locked into technologies untouched by the modern world. It was instead a forward-looking attempt to forge dynamic alternatives to the lifestyles currently available in the larger settlements of central Australia. It was pointed out at TAGAL's 1980 workshop that the 'movement to outstations does not mean a return to a traditional existence, and it is therefore necessary to introduce some modern technology [such as vehicles, communication devices, bores and rifles] to achieve ends which are no longer achievable by traditional means'.[101] Walker's thinking, however, went a stage further: that the technologies to be introduced needed to be tailored to the users' needs and aspirations rather than imposed upon them.

This was not an easy task. Walker's thinking about appropriate technology generated two immediate frictions as he began to apply them in Aboriginal communities during the 1980s: some residents responded explicitly that they wanted whitefella technologies rather than adaptations of them, and many more held back from participating in community training and construction activities as the new facilities were installed. CAT plainly had to rethink the fundamentals of its mission. Walker conceded in 1988 that 'Community support for [CAT's appropriate technology] programme has taken a long time to develop. Many people are sceptical and feel they are getting second class options. Unfortunately people have been educated to think that 20th century technology has all the answers and is the only way forward'.[102] It became quickly evident that outstation residents, by opting to live away from larger communities, had not necessarily chosen to forgo all the facilities and activities that would have been available to them. During a visit by CAT engineers to Mantamaru (Jameson), near Warburton in Western Australia in 1983, Charlie Munro told the visitors that the community 'wanted silver button/push button toilets that fitted into their houses just like the one in the [Council] visitors quarters'.[103] CAT's VIP pit latrine seemed a poor alternative. With housing, likewise, one CAT fieldworker conceded in 1990 that the 'current expressed desire at Walungurru [is] to live just like the whitefellas – to occupy dwellings like those lived in by community advisers, storekeepers, teachers and so on'.[104]

CAT staff were also dismayed that outstation residents did not always participate enthusiastically in deploying the new facilities. They perhaps did not allow sufficiently for the sense of frustration and disengagement among local residents because key decisions – however well-intentioned – about the facilities to be provided had already been taken, and areas of competency in their installation had already been mapped out for them. Walker experienced just such a situation in 1984 during construction work in Harts Range north-east of Alice Springs. Three local men had been nominated to assist the CAT team, but Walker reported back to the community 'We have not had a very successful week'. He explained that 'Despite our coaxing [the three local men] appeared reluctant to join in with the work and drifted off on the first afternoon only to appear drunk late that afternoon and suitably hungover the next morning. This practice has been the case each day since'.[105]

Plainly something had gone wrong, and Walker wanted to understand the three men's side so that CAT could engage more successfully with the community in the future. Several months later he did get the other side of story, when the Atitjere Aboriginal Community in Harts Range contacted a wide cross-section of agencies and organisations to say that the community wanted better opportunities to express its own views about 'future developments for our community'. They wanted to make clear 'what kind of shelters we want, where we want them built, and what improvements we can have made to them', including water supply, showers, laundries and toilets. The community made it especially clear that 'We want to talk to all organisations together because we have visits from different organisations at different times and we hear different stories so that we don't know what is really happening. We don't want to have so many meetings and never know what is going to happen'.[106]

CAT's response to such tensions was threefold. Firstly, it undertook community development work only if asked to do so by local community representatives, and for only so long as they retained a community's confidence. For example, the Kaltukatjara Community Council in 1985 invited CAT to Docker River, explaining that 'The community ... would like to consider forward planning with more independence in mind as a goal'.[107] The Waringarri Aboriginal Corporation at Kununurra likewise contacted CAT in 1988 to 'get some good ideas for developing our Out-stations'.[108] In partnering with such communities, CAT took advantage of the Australian government's new Community Development Employment Projects Scheme (CDEP), which had begun in 1977 and was expanded in 1987. CDEP's purpose was to translate individual unemployment benefits into accumulated grants that community councils could use to employ the locally unemployed in community development activities. CAT sought to harness CDEP to its off-campus technical training activities in order to maximise community choices, participation and training. At the conclusion of its shelter-building programme for Walungurru's outstations in 1990, for example, CAT staffer Kurt Seemann noted with satisfaction that the 'occupants and assistant builders often express how proud they are for having been part of the project from start to finish'.[109] CAT's ATWORK programme grew out of these CDEP partnerships, and was intended to further empower Aboriginal people to control the selection and use of technologies appropriate to their settlements. Describing ATWORK's early achievements, Walker pointed out in 1989 that a 'significant number of beneficiaries of CAT's initiatives are women. The irony is that while the majority of people in remote communities are women, they are under represented in technical programmes. We are concerned that women are being excluded from technological decision making and are being excluded from consideration in technological issues'.[110]

Secondly, CAT emphasised that appropriate technologies should not be seen in a vacuum, but as elements within wider environmental contexts and socio-cultural processes. As Walker had remarked in 1980, 'The imposition of hardware (i.e. the physical solutions of needs – trucks, tractors, houses, machines, etc) is not of itself a sign of development. The software component must be present before & after to ensure success'.[111] By 'software' Walker already had in mind the cultural emphases and resulting lifestyle choices of the participating communities. By the end of the decade, CAT had accumulated the research capacity to be able to begin to explore how these processes played out in remote settlements, especially in relation to choices about waste management and solar distillation of water. By 1989 its waste management solutions had attracted interest across Australia, and won funding support from the Northern Territory Department of Health and Community Services.

Thirdly, in paying attention to the processes and systems within which technologies operated, CAT began to advocate an explicitly holistic approach to making decisions about life choices in remote communities. This approach was based in part on Walker's earlier training in ekistics, but it was also nurtured by his increasing familiarity with Aboriginal knowledge systems. He argued that appropriate technology was 'a people centred approach to development' that looked beyond material measurements of improvements in living standards to focus instead on the 'holistic concept' of happiness, safety, and 'overall quality of life'.[112] CAT drew attention to objections by Aboriginal women that, for example, cultural and spiritual elements were often overlooked in community design, and argued for recognition of Aboriginal concepts of science and technology that resonated with their lifestyle and environment.

Walker became guardedly optimistic that as a result of gradually more sophisticated community development approaches during the 1980s there 'began to emerge … a [wider] realisation that there was an Aboriginal identity which required a different technological value base if it was to survive'.[113] However, he conceded that many Aborigines did not accept his reasoning and expected that city lifestyles could easily be translated into the Outback. Walker began to talk about the 'stress caused by the dual goals' of living on country while wanting the same facilities and services as were available in towns and cities.[114] Walker also recognised that in attempting to resolve these tensions the mindset of the broader Australian community was not helpful:

> We pay lip service to, but don't seriously consider, the technologies of Aboriginal people and how these technologies kept them alive and healthy ... I am not indicating a return to past traditional practices. I am trying to draw together an argument which says there may be a case for a different arrangement of values and lifestyle and hence technologies and infrastructure and medical standards and guidelines which could present and preserve Aboriginal culture and lifestyle in a much more enduring way ... I say this because I believe progress along the western development model only preserves Aboriginal dependence on the wider Australian community and despite the rhetoric of self determination, when it comes to the technologies which support Aboriginal decisions the power and control is outside the community.[115]

As CAT pointed out the disconnections between Aboriginal community aspirations and public policy outcomes, it also cautiously suggested that although it was morally difficult to argue against the principle of universal human rights when servicing remote communities, the effect of attempts to provide universal entitlements to goods and services had been to entrench Aboriginal poverty and dependence. Walker began to think – although his arguments were complicated by his strong commitment to the right of all Aboriginal communities to access safe and reliable water supplies – that policy thinking should focus on ensuring livelihoods rather than catering to rights. He contended that without effective wealth creation activities, living in remote communities 'has the drawback of never breaking the dependency of Aboriginal people on government nor does it give them the power to lobby for increased benefits'.[116] In 1991, reviewing a decade of CAT's experiences working with homelands communities, Walker said that in the absence of 'aboriginal and islander control and self-determination and the input of aboriginal and islander values to technical decision making[,] little substantive progress will be made in areas of service provision which attract large technological intervention, such as water and sewerage, public health and housing'.[117] It was an astute but depressing message.

Chapter 3

Realignments: the 1990s

The 1990s was a decade of radically changing fortunes for Aboriginal Australians, and of fundamental realignments at CAT. Nationally, the Aboriginal and Torres Strait Islander Commission (ATSIC) was established in 1990 by the Australian government to represent all Aboriginal Australians. It comprised two elements, one representative and the other administrative. The first was made up of elected regional councils from across Australia, representatives from which in turn elected a national board of ATSIC commissioners. The second comprised federal public servants who implemented decisions made by the commissioners. ATSIC took over from the Department of Aboriginal Affairs the administration of programmes for Aboriginal and Torres Strait Islander people. In a related local initiative designed to promote Aboriginal decision making, CAT established in Alice Springs an Aboriginal board of management, comprising representatives from ATSIC regional councils and chaired by Jim Bray, to oversee the organisation's activities. Nationally, the parameters of Aboriginal engagement with the Australian state were redrawn by the 1992 Mabo and 1996 Wik High Court rulings, which recognised enduring native title rights to traditionally held homelands. These rulings (and in parallel with them, CAT's water reports for the Human Rights and Equal Opportunity Commission in 1994 and 1999) were the culmination of the human rights catch-up approach to Aboriginal policy making that had developed during the 1970s and 1980s, but which during the 1990s was gradually translated into a modified strategy that focused on the provision of appropriate entitlements for remote Aboriginal communities.

In a less positive vein, Aboriginal affairs during the 1990s were increasingly conditioned by a new tone of 'conservative neo-liberalism' in Australian public policy debate. Conservative neoliberalism is essentially 'a "negative" Liberalism concerned to get the state out of markets, deregulate and privatize, [and] reduce social and bureaucratic spending'.[1] This was highlighted federally by the free-market and mainstreaming approaches of Labor governments under prime ministers Bob Hawke and Paul Keating (from 1991), and of the Liberal-National Party coalition government under John Howard from 1996. Neoliberal principles were evident in mounting criticisms of the human rights agenda hitherto pursued by the Australian government for having delivered only 'passive welfare' that intensified Aboriginal dependence, and by negative stereotypes of Aboriginal homelands as marginal and unsustainable places that were a drain upon taxpayers' dollars. CAT had emerged and consolidated itself during the 1980s on the basis of bipartisan policy support for the homelands movement and of strong local partnerships that could harness the energetic participation of those communities in their future development. Neither of these elements could be relied on by the later 1990s.

Although the criticisms of passive welfare were exaggerated, it was true that the increasing governmental red tape surrounding the servicing of Aboriginal communities tended to stifle local decision making and willing participation in community development activities, and thus constrained the community partnerships upon which CAT's activities had been built. Such tensions were symptomatic of a fundamental shift in policy away from the loose consensus on Aboriginal self-determination and direct support from the public sector for Aboriginal wellbeing that had characterised national affairs during the Whitlam and Fraser governments. During Hawke's prime ministership, and also after Keating replaced Hawke as leader, federal governments set about restructuring the Australian state and economy in order to stimulate private sector growth. Direct public sector engagement in service delivery was translated increasingly into bureaucratic oversight of services provided under contract by private service providers, an approach that was reinforced after Howard won the federal election in 1996. These changes had the effect of winding back Aboriginal autonomy, despite Aboriginal empowerment initiatives promoted by both the Hawke and Keating governments, and Howard's 'practical reconciliation' strategy to reduce disadvantage in 1998, as jobs in the rural sector and the informal economy were lost and as the mainstreaming of services for Aboriginal Australians resulted in a renewed 'colonisation' of Aboriginal communities by non-Aboriginal bureaucrats and consultants.[2] The gradual reformulation of the Australian state and economy, and of the emphases of Australian public opinion, were symptomatic of a broader switch across the full spectrum of mainstream political opinion in the Western world towards free market capitalism, a change which has been loosely called neoliberalism and economic rationalism.

As these policy changes took tangible form in local affairs during the

1990s, there emerged within the Aboriginal homelands disturbing signs of frustration, tension and internal division in reaction to increasing state-sponsored interventions in their communities. This became evident in even the most purposeful and committed of homeland settlements. At Walungurru (Kintore), for example, in the heartland of the Northern Territory-Western Australian border region where the Western Desert art movement began (and where CAT's first hand pumps were installed in the early 1980s), local spokesman Andrew Spencer complained in 1997 that 'W[hite] F[ella] come here and just drive around and we don't know what they are doing here – they then make decisions'. Artist Cameron Tjapaltjarri added, 'We don't know Gov side – always changing'.[3] Such attitudes were accompanied in some communities by disengagement, despair and violence. At Walungurru, which officialdom increasingly came to regard as 'one of the more difficult communities to which government has to respond',[4] elder Dixon Fisher lamented in 1997 that 'families have problems as kids grow older – [petrol] sniffing, drinking, stealing & fighting'. Peter Sutton has argued that

> The Achilles heel of post-1970 Indigenous policy history was that the new levels of unemployment, new freedoms from authoritarian control, new concentrations of populations formerly dispersed, new accessibility of drugs, new alcohol-purchasing power, and new 'sit-down money' for the unemployed, were not matched with measures designed to assist people through the crises of occupation, discipline, motivation, conflict management and community trauma that soon erupted and by the 1990s had reached a crescendo, especially in the remoter regions.[5]

* * * * * * * * * * * * * * *

The 1990s were also a time of fundamental realignment for CAT, which in 1995 formally separated from the Technical and Further Education (TAFE) College in Alice Springs where it had begun. The decade was a period of massive and positive achievement for CAT in terms of the geographical scope of its operations, revenue, staff numbers, research innovation and national influence. These transformations were recognised in 1999 when CAT's director, Bruce Walker, won the prestigious annual Clunies Ross National Science and Technology Award.

CAT's activities extended far beyond the Alice Springs region where they had begun in the 1980s, with the result that Queensland and Western Australian offices were opened in Cairns in 1993 and Derby in 1998. CAT's Aboriginal technical training programme (ATWORK), begun in the late 1980s, became a national undertaking during the early 1990s. Simultaneously, from the late 1980s until the mid 1990s, Walker prepared an Australia-wide *Water Report* for the Federal Race Discrimination Commissioner on water provision to Aboriginal and Torres Strait Islander communities. In 1994 CAT established a trail-blazing National Technical Resource Centre (NTRC) with nationwide responsibilities, a wide-circulation periodical (*Our Place* magazine) and an expert national advisory committee, which further broadened CAT's horizons and the range of its activities. A pioneering national conference on Aboriginal women and technology was held in 1996. Kurt Seemann, who was intimately involved in launching both ATWORK and the NTRC, recalls that 'a burst of very novel new strategies seemed to flow from our work' during these years.[6] ATWORK, the water report and the NTRC are discussed in greater detail later in this chapter.

Through these innovations there also developed a relationship of the utmost importance for CAT's future: Peter Taylor, a senior federal manager in ATSIC, increasingly looked to CAT to deliver outcomes to remote Aboriginal communities that he doubted could be achieved through mainstream avenues. Walker says that Taylor effectively became 'our funder'.[7] Early in the next decade their collaboration would culminate in Bushlight, a revolutionary new solar energy supply system for remote Aboriginal communities and, when CAT moved to the Desert Knowledge Precinct in 2010, Walker handed over leadership and Taylor replaced him.

CAT developed other important partnerships during the late 1990s. In 1997 it became a partner in the new Australian Centre for Renewable Energy (ACRE), one of a network of cooperative research centres funded by the Australian government. This alliance would supply the groundwork for what later would become Bushlight. In 1996 CAT began to discuss collaborative activities with the mining giant Rio Tinto, and as this partnership matured during the decade it was formalised by the signing of a memorandum of understanding in Parliament House, Canberra, in 1999.

Parallel to these developments, CAT's staff doubled to about 40 during the 1990s. CAT's annual operational budget increased from around $1 million early in the decade to some $3.5 million by its close. As income increased, its sources changed. Sales of CAT homelands products, which had resulted in the centre's expanding activities and growing independence from the college during the 1980s, declined from almost a third of CAT's total annual budget to an almost negligible level, whereas ATSIC funding rose from around 10 per cent to 27 per cent, and training related income (dependent upon, as was ATSIC's funding, the federal government) increased from 47 per cent to 56 per cent. CAT had seemingly exchanged its initial dependence upon the local TAFE college for a new dependence – riskier because the stakes were so much higher – on the Australian government. As Walker acknowledged late in the decade,

> the biggest barrier CAT found to doing things is the is sue of management and the availability of resources … rather than the technical challenge that is involved. CAT will always have to get 7 different funding agencies together in order to build a house if policy issues aren't addressed.[8]

CAT's growth came with other associated costs. Production at the Priest Street workshop, which had been the hub of CAT's activities during the 1980s, and which had first established CAT's reputation in the homelands, slumped after the mid 1990s. The workshop was a victim of changing government policies, funding formulas and procedures relating to Aboriginal homelands and their servicing by the private sector. Andrew Lane, an Aboriginal architect who managed the workshop and sat on the CAT board until 2004, calculated that it took three days to make a drum oven, and the cost of wages, materials and freight made it difficult to sell.[9] ATWORK too, notwithstanding its innovative flair, would suffer because of its dependence on national funding mechanisms and accountability to norms set by the mainstream Vocational Education and Training (VET) sector. Substantial ATSIC funding for CAT, and the rigorous accountability that this entailed, led to internal anxieties about the robustness of CAT processes and the enduring strength of the organisation's appropriate technology goals. The transition in management and governance from what had been a one-man show in the 1980s to a large organisation with decision making shared between its founding director, an Aboriginal chairman and board of management, senior executive officers and general staff meetings (representing a wide variety of work specialisations), led inevitably to tensions and soul searching about the organisation's structure and ongoing relevance. As one widely respected CAT manager recalled, CAT became a 'bull pit' during these years.[10] These anxieties, together with the broader tensions caused by changing government policies, led Walker in 1996 to circulate a discussion paper, 'In Twenty Years Time', which warned that for all CAT's apparent success it remained, like other Aboriginal sector organisations, 'largely dependent on the whims of government for our survival'.[11] Walker challenged the organisation to reinvent itself in order to confront an uncertain future.

1. CAT Inc.

Although CAT's operations became financially largely independent of the Northern Territory TAFE sector during the late 1980s, it was not until the

Jim Bray and Bruce Walker display CAT's new certificate of incorporation, 1989.

1990s that CAT achieved full organisational independence and Aboriginal oversight. Its contributions during the 1980s to the central Australian region around Alice Springs broadened during the 1990s as it developed a national presence, underlined by Walker's water report, and underwritten by large new funding agreements, most notably with ATSIC. These transformations were accompanied by visible marks of success at CAT's Priest Street headquarters in Alice Springs: new buildings (federally funded) were opened in 1993, and in 1998 (enabled by the Northern Territory government) CAT obtained freehold title to the Priest Street property. Behind this public face of effortless achievement, major tensions and uncertainties developed in CAT during the 1990s.

Its incorporation in 1989 and increasing organisational independence – and thus its apparently unfettered scope to assist 'remote' communities – masked the realities of ongoing constraint within the Northern Territory TAFE sector until mid decade, and of continuing and increasing dependence throughout the decade upon policy makers in the national and state capital cities, remote from the Aboriginal communities with which CAT worked. Even within its internal core, tensions surfaced about decision-making processes, staffing and structural organisation. These tensions are explored in the following pages.

One staffer remarked in 1995 – the year in which CAT became a fully autonomous organisation and embarked upon a major new funding relationship with ATSIC – that low staff morale and uncertainty about the future were directly attributable to CAT's apparently headlong successes: 'our programs and services are growing so fast', he remarked, but 'we haven't grown inside CAT itself to match that growth in the outside services that we provide'.[12] However, other staff relished this work environment. For Mark Moran, who established CAT's operations in Queensland during the 1990s, one of the most 'remarkable thing[s] about CAT is how flat the structure is', without continuous direction and micro-management from above, so that 'you actually had a lot of freedom' to innovate.[13] However, whether one felt out of one's depth or relished the opportunity to test the boundaries of possibility, all CAT staff 'felt very much under the pump in the mid 1990s'.[14]

It took a major outside review by Doug Porter to sort out the mixed thinking within CAT. As Walker later admitted, until the Porter review CAT had achieved a large measure of operational, but not organisational, maturity: 'We had national funding, and raced off doing random projects but we didn't have any strategic intent'.[15]

****

CAT's relationship with the Northern Territory TAFE sector ended very much as it had begun: amid controversy and bickering. CAT's formal incorporation took effect in November 1989, but it remained a unit within the Alice Springs College of TAFE (ASCOT). A memorandum of understanding between the two bodies, with the chairmen from both organisations invited to join each other's governing bodies, did little to improve the suspicions that had clouded the relationship during much of the 1980s. On the one side CAT sought to maximise its independence, and on the other side college management sought to maintain oversight. CAT's success in winning large-scale external funding for applied research projects of national significance put the college's small-scale management and accounting processes under enormous strain. Walker's water project first revealed these incompatibilities, Walker complaining that college red tape meant he had to 'work in a straitjacket'.[16] CAT's subsequent funding relationship with ATSIC escalated these tensions. The chairman of ASCOT's council, Dr Richard Lim, complained in 1992 that 'the Council needed to be more informed about what CAT was doing … ATSIC were about to hand over approximately $1.5 million [to CAT], and the Council wanted to know more about how this money will be spent'.[17] Lim suggested to Walker that a joint working party be established 'to resolve outstanding issues and misunderstandings which have caused uncertain working relationships between CAT and the College and to determine a future partnership'.[18] Walker replied coolly that the college was kept sufficiently informed of CAT's activities, and that 'the CAT Board should be treated as a separate but equal entity'.[19] Although he undertook to review and improve CAT's reporting processes, he was not prepared to compromise its autonomy by turning CAT into 'a sub-committee of ASCOT Council'.[20] Privately he said of the college that CAT's operations were 'no business of theirs'.[21] From Walker's perspective the college had always disdained CAT's activities because of 'an unwritten negative view of Aboriginal Programmes', but with CAT now 'turning over about $1million a year and [likely to] receive substantial support from the Federal Government, the Chair and new members of the College Council are suddenly wanting to have a say in how CAT's operations are progressing'.[22]

Relations with the college deteriorated further after ASCOT was restructured as Centralian College in 1993 upon its merger with Sadadeen Senior Secondary College. The college's new director, Diane McEwan, disliked the terms of the existing relationship with CAT. She felt that she rather than Lim should sit on CAT's board, and she was uncomfortable about CAT funding an increasing number of staff positions over which she had no clear control. McEwan sought to forge a new partnership, the intention of which Walker interpreted as winding back the clock to make 'the Director of CAT [a] part of the Centralian College, with little independence' and unambiguously subordinate to the college director.[23] Bray also objected to what he saw as the college's efforts 'to dominate' CAT, and said that as a result the relationship between the two 'has soured a lot'.[24] Bray fretted that CAT might lose its remaining operating funding from the college if it separated any further from it, but Walker was less concerned about this than with retaining control over the substantial reserves that CAT was building up. He warned CAT's board that although the organisation had accumulated some $773,000, he had no signing authority over these funds and depended entirely upon 'the understanding between C.A.T. and the College that all funds earned … through entrepreneurial activities will remain under the control of C.A.T.'.[25] Walker sought to bypass the college altogether and to develop a direct relationship with the Northern Territory government. Some of these exchanges occurred in flannels rather than business suits: Walker was a keen cricketer, and 'very high level "negotiations" … occurred between Bruce and Government folk on the cricket pitch between … stumps!'.[26] Walker finally achieved full autonomy from the college when he won an independent funding agreement from the Northern Territory education authorities late in 1994. This was formalised on 1 July 1995, from which time CAT Inc became the employer and account holder for all CAT operations.

* * * * * * * * * * * * * * *

Tensions over issues of governance were not limited to the complex relationship between CAT and the college. CAT's incorporation in 1989 entailed creation of a constitution. This was modelled upon that of CAT's sister organisation in Alice Springs, the Institute for Aboriginal Development (IAD), which required that the organisation's director report to an Aboriginal board of management comprising representatives from the new ATSIC regional councils in central Australia. An interim board was established in November 1990 with Bray – who was widely respected for his work as manager of CAT's residential complex – as its chairman, pending the full establishment of the regional councils. However, because of the time needed to create a fully functioning national ATSIC network, it was not until August 1992 that the first meeting of CAT's permanent board, and its first annual general meeting, was held. During the uncertain interregnum from 1990 until then, misunderstandings inevitably arose over transitional arrangements about decision making, information sharing and ultimate control.

Tensions came to a head at a meeting of the college council in 1992, when some of its members, concerned that they knew so little about CAT's operations and that Bray seemed equally in the dark, objected 'that James as Interim Chairperson of CAT Inc wasn't being kept informed'.[27] Bray, embarrassed and upset, summoned a special meeting of CAT staff to determine how the organisation was really going to be run. At that meeting Walker for the first time fully described to his colleagues the difficult working environment that had characterised CAT's relationship with the college since his arrival in 1980. Bray was partially mollified, but responded that had he been fully briefed before college council meetings he would have been able to respond better on CAT's behalf 'when difficult questions were asked'.[28] Bray reiterated his dissatisfaction later in the year, when he remarked that notwithstanding the semblance of a switch in decision making from CAT's director to the board, he remained 'Concerned about the real power of the Board'. He asked, 'What will [the board] really do?' and demanded 'clarification of the board's role as at the moment the old system is still in place'.[29] The relationship between Bray and Walker would never become 'cosy' and was occasionally punctured by 'heated argument', but it was always 'respectful' and effective, and often coloured by humour. Walker recalls that 'Jim's resigned a number of times; he's sacked me a number of times'.[30]

Bray's concerns about internal governance gradually diminished after 1992 as CAT's board of management took permanent form and established effective working relationships with Walker and senior staff. However, as these relationships firmed, new tensions arose between the senior management group and general staff. Some staff felt that there was 'a "them and us" situation with the Board', and others likened the latter to 'a "Grandfather" who won't acknowledge the skills, intelligence and opinions of the younger Aboriginal staff'.[31] CAT's apparent 'Aboriginalisation' policy, which the board seemingly personified and which it reinforced by deciding in 1996 to employ an Aboriginal CAT executive officer, also caused disquiet amongst staff. Some Aboriginal employees queried whether the board had pushed the policy far enough: 'CAT's Aboriginalisation policy needs clarification. Does it exist? If so where is it? If not, why not? Is it thorough enough?'. Non-Aboriginal staff worried about job security, one of them predicting that 'because I am not Indigenous, I will be seen as just another whitefella educated idiot'.[32] In response the board announced that it 'has consciously chosen not to fully aboriginalise ... CAT in order to foster greater levels of understanding between aboriginal and non aboriginal people'.[33] Another staffing issue emerged as CAT expanded and more clearly articulated the philosophy and priorities that underpinned its operations, because as Porter observed, 'it took far longer than anticipated to orient both existing and incoming staff to the new corporate vision and mission'.[34]

* * * * * * * * * * * * * * *

These uncertainties about the character and future direction of CAT resulted in a period of sustained introspection from the middle to the end of the decade, and these led to internal reviews and reorganisations in 1995–96, 1997 and again in 1999. The first of these reassessments originated in 1995 as CAT ended its association with Centralian College and responded to suggestions from its new NTRC national advisory committee about future directions and capacity building. As a result, in March 1996 a new organisational structure was unveiled, the main elements of which were the NTRC, a Research, Design and Planning Unit, an Education and Training Unit, a Development Services Unit and a central executive, together with a mission statement and strategic plan for the rest of the decade. In December 1996 these recalibrations were confirmed when a revised constitution was approved, another effect of which was to allow CAT to be registered as a public benevolent institution in order to save hundreds of thousands of dollars in tax concessions.

These readjustments were coloured by two parallel developments, which led to a second round of more far-reaching discussions and reforms during 1997. The first input was Walker's 'In Twenty Years Time' paper, which he presented to the board in June 1996. It would influence debate for the remainder of the decade and beyond. Walker argued that CAT and other Aboriginal organisations in central Australia would need to reinvent themselves and develop a bolder shared vision if they wanted to retain relevance over the next 20 years. He cautioned that they all risked being 'marginalised in the back streets of Alice Springs', and – pointing to the overlapping activities of organisations such CAT, IAD, the Central Land Council, the Tangentyere Council, and Batchelor and Centralian Colleges – he argued that 'While most organisations in Alice Springs have survived, none of us appear to have prospered. All of us have been competing for newspaper space, funds, resources and clients'.[35] Walker saw that growth had resulted in CAT now being 'land locked' at Priest Street and needing to rebuild or relocate in order to increase its public profile and sustain its expansion. He suggested that at least some organisations should relocate onto a shared campus in order 'to develop a physical presence that will establish a critical mass of support to lead us forward in new fresh opportunities that support sustainable communities'.[36]

The second input came from community development expert and public sector analyst Doug Porter, who from 1995 to 1997 comprehensively reviewed CAT on behalf of ATSIC. Although Porter's review was, strictly speaking, confined to the workings of the NTRC, which had been funded by ATSIC, the NTRC was so thoroughly integrated into CAT's overall activities that the review amounted to an assessment of CAT itself. Porter interacted closely with the NTRC's national advisory committee, and CAT management and staff during 1995–96, and continued to visit CAT during the following year, delivering an interim report in February and a full draft report in July so as to stimulate discussion within the organisation, and ultimately a final report in September 1997.

Porter's comments and recommendations resulted in a comprehensive 'tune up' of CAT's structure and processes, which were approved by the board in December 1997.[37] Porter had queried whether there was sufficient clarity of focus behind CAT's evolving activities, suggesting that 'there was confusion within CAT and outside CAT as to the extent and purpose of the NTRC as distinct from CAT. A centre within a centre is a difficult concept to grasp'.[38] More broadly, he argued that the successful development of the NTRC was dependent upon fundamental 'institutional change, in particular, completing and giving effect to the strategic planning process, a better sense of direction and managerial responsibility … and giving greater attention to staff development'.[39] CAT responded by restructuring its diverse activities into two divisions, each headed by an Associate Director: Corporate and

Technical Services (which brought together the NTRC – renamed the National Technical Services Clearinghouse at the end of the decade – and CAT's other applied research activities), and Program Services (which included CAT's VET programmes and its extension services). In addition, it was decided that CAT's executive would be assisted by a management team that would oversee the implementation of executive decisions, directly mentor staff and act as 'a forum for CAT values and views'.[40]

CAT's activities were further assessed in 1999. It was prompted in large part by internal concerns that CAT, although it had benefited enormously from the funding received from ATSIC during the decade, was becoming too reliant on ATSIC for its ongoing development. This close relationship was seen to carry the financial risk of over dependence upon one funding stream. It also seen to carry the operational risk of limiting much of CAT's attention to activities that ATSIC was prepared to fund. Members of the board worried that CAT was not necessarily addressing the requirements of their constituents in remote Aboriginal communities. Bray expressed concern at a meeting of the national advisory committee in November 1999 'that Aboriginal people do not benefit to the full extent from work undertaken by CAT'.[41] The board held a two-day workshop in December to review these issues and identify a better way forward, and drew again upon Walker's 'In Twenty Years Time' paper which suggested that identifying ways 'to derive independent income is probably the only opportunity we have to break the dependency on Government support'.[42] The ideas canvassed at the workshop were radical: CAT should aim to generate at least a quarter of its operational budget from commercial activities; it should renew the board's membership and plan for a transition from its foundation director; it should modify the 'rights based' approach to its community development activities; and in collaborating with other organisations to establish a one-campus Desert Peoples Centre (DPC) in Alice Springs it should think about 'more than "Education & Training"' and take the lead in 'calling the mob together [and] setting the agenda' for wider initiatives.[43] Some wondered whether CAT was 'slowing down' in its community development activities and was not providing 'enough empowerment through the current method of service delivery'. There was broad agreement that CAT 'needed to be more out there'.[44]

2. Changing Operations

Late in 1996, in the midst of internal debate about organisational structure and processes, Walker confided to CAT's new executive officer, Jenny Kroker,

I want to focus next year on starting to look outside CAT, to develop a sense of purpose and urgency on what is happening in remote communities. Less internal discussion and refinement on how we might organise ourselves, our conditions, whether we should have this computer or that and more on the external drivers to CAT.

We have less time for internal soul searching.[45]

Walker's sense of urgency was influenced not only by the altering policy environment that was being created by Canberra, but also by the radically changing nature of CAT's engagement with remote communities, which is explored in the following pages. During the 1990s CAT's activities evolved from technical design and several small training courses inherited from the college, reflecting how the centre had operated during the 1980s, to design work integrated with major programmes in research and technical education. This more holistic approach was accompanied by a broadening of vision from regional to national engagement. The three catalysts for change were the preparation of Walker's national water report, the launch of ATWORK and the establishment of the NTRC, all of which are described in the following sections. In 1992 the first issue of the *ATWORK Newsletter* signalled the changed orientation of CAT's activities, describing the organisation as 'an Aboriginal research, design and education organisation in Alice Springs'.[46] Bray drew attention to the effects of this reorientation in 1997, when he announced that the NTRC had enabled CAT to 'expand from an operation based largely in Central Australia to an organisation with a broad national outlook'.[47] The NTRC immediately embarked upon a wide range of cutting-edge, applied research activities in community development, including projects on bore water quality, hot water systems and washing machines suitable for remote communities, and research on shelter design, renewable energy and waste management.

These three catalysts for change created a new operational environment that in turn made possible other key CAT initiatives during the 1990s. These included community planning (highlighted by Old Mapoon, 1995), local governance capacity building (for example, Walungurru, 1997–98), and the forging of important alliances with other organisations (for example, the partnerships with ACRE and Rio Tinto). The publicity surrounding CAT's new steps spruiked success: the formal sign-off on funding for the NTRC in October 1994 was attended by ATSIC commissioners Charles Perkins and John Patterson and received national media coverage. So too did the signing of the partnership with Rio Tinto in Parliament House in 1999, in the presence of the Chief Scientist of Australia and the Minister for Aboriginal and Torres

Strait Islander Affairs. But the silence in the old Priest Street workshop as CAT's production role diminished after 1996 led some to question how CAT's new national role would affect its ability to connect with remote Aboriginal communities.

* * * * * * * * * * * * * * *

CAT's continuing relevance depended upon the ongoing strength of its local relationships with homelands communities, notwithstanding the magnitude of change within both those communities and CAT during the 1990s. In 1995, for example, CAT staffers Sonja Peter and Christian Tietz travelled to the small Aboriginal community of Kaltukatjara (Docker River), in the far south-west corner of the Northern Territory, to deliver a CAT drum oven that had been ordered from the Priest Street workshop by the Kaltukatjara Women's Centre. It was a routine job. While demonstrating how the oven operated they chatted with local women about their cooking and storage-facility needs: 'The women said that their electric oven and stoves as well as their refrigerators were all broken down and that none of them had storage facilities. During the discussion the difficulty with electricity emerged and the preference for timber burning options was voiced'.[48]

This was the same sort of conversation that Walker had engaged in during the 1980s, and which had set his early goals for CAT. The organisation's activities switched during the 1990s from the production of simple homeland infrastructure to more complex activities in community development for remote settlements, such as the construction and maintenance of water services, sanitation systems, buildings and roads, together with associated training: for example, on-site instruction in power tools, welding, building design and construction, waste reduction and maintenance of solar hot water systems and vehicles. These services were provided across a widening area, from Papunya north-west of Alice Springs, to Daly River, south-west of Darwin, Ngukurr in south-east Arnhem Land; Leonora, a mixed race mining town to the north of Kalgoorlie-Boulder in Western Australia, and Mulga Bore, an outstation within the Anangu Pitjantjatjara homelands that straddle the South Australian-Northern Territory border. CAT's evolving community-development activities also resulted in community design work and planning, for example at Old Mapoon on the western Cape York peninsula, and project management, such as at the Doomadgee outstations in Queensland.

Notwithstanding the widening geographical scope and complexity of these operations, CAT was forced by changing government policies and programmes into a juggling act with multiple grant agencies in diverse tiers of government in order to maintain the level of community engagement that it had established during the 1980s. At Walungurru and surrounding outstations, for example, a fluid community of 300–500 people in the homelands region north-west of Alice Springs where CAT's activities had begun in the early 1980s, CAT applied Australian government funding for the development of its new ATWORK Aboriginal training programme during the early 1990s in order to run a pilot project that combined vocational training with community development. Seemann became the project coordinator, overseeing trainees in house repair and vehicle maintenance. Rick Callahan was employed as building manager, and in the process he also provided technical training and practical help with housing, water supply, sanitation and solar power to outstations. Callahan, a New Zealand builder, had become the backbone of CAT's extension services during the 1980s, and 'dug holes and poured concrete across most of remote Australia'.[49] Clarrie Maher, a West Australian Aboriginal diesel mechanic who had completed an undergraduate teaching degree while working at CAT, lived at Kintore alongside Callahan and passed on automotive and construction skills to the local work crew. In 1991 ABC television's 'A Big Country' showcased the official opening of a residential complex and workshop at Walungurru that had been built by local CAT trainees under Callahan's supervision. During the ceremony Smithy Zimran, chairman of the Walungurru community council (WCC), reminisced fondly about the community's long partnership with CAT:

> Bruce Walker first started coming out here eight years ago when we began this place from scratch. You will all remember those days and how CAT helped us with a handpump and then some hand washing machines long before the Government became interested in Kintore. Steve Patman from CAT worked here for 2 years on our first houses. Later CAT built the VIP toilets and shower blocks which are still over there and after that helped the school with more toilets.
>
> Handpumps on all the outstations and the Ngurra Wayltja outstation building program came next and that was when we began to get to know Rick pretty well. He helped us learn about building our own places.[50]

In 1994 there was further good news when the then leader of the federal Opposition, Alexander Downer, and Brendan Nelson, who was then president of the Australian Medical Association (he was elected as a federal Liberal MP in 1996 and would later become party leader), were shown over four new houses at Walungurru that had been built by CAT staff and trainees.

However, difficulties had quickly emerged over how CAT's initiatives would be sustained. Outside consultants and contractors were increasingly

Clarrie Maher instructing students in Automotive and Mechanical Skills.

awarded local works projects despite the community's expressed desire for greater control. As CAT's direct funding through ATWORK for community participation and training in local construction and maintenance work was exhausted, Callahan redirected residual funding from other CAT grants in order to continue local operations. As he put it, 'We have been able to survive here at Kintore by robbing Peter to pay Paul, [and] this has kept the present maintenance program going'.[51] However, in 1995 CAT ceased its operations at Walungurru because it could no longer continue to cross-subsidise local work from other sources. As Walker admitted, 'things [thereafter] fell in a hole', with many in the community 'feel[ing] like they have been abandoned', because the WCC (created in 1983) was unable to fill the void left by CAT's departure.[52] The community council was hampered by limited funding and high staff turnover, and struggled to broker a consensus between the emerging 'generation cultures' within the community.[53] Houses deteriorated, drains blocked, water pipes leaked and solar water heaters broke down.

Difficulties in funding local community development work were likewise evident during the beginnings of CAT's operations in north Queensland and the Kimberleys in Western Australia. Federal funding through the Community Development Employment Projects Scheme (CDEP), which was introduced to remote Aboriginal communities from the late 1980s, and which became their mainstay for community development activities during the 1990s, opened up some possibilities for CAT. In 1994, for example, it was awarded funding through the Doomadgee Aboriginal Community Council to manage outstation housing and infrastructure for a cluster of outstations with a population of some 1,000 people around the Old Doomadgee mission site, about 500 kilometres by road north of Mt Isa, where the Ganggalida people had established outstations since the early 1980s (Callahan was redeployed there when CAT's operations in Walungurru were suspended in the mid 1990s). In 1995 CAT drew upon CDEP funding to the Aurukun outstations on Cape York Peninsula in far north Queensland to oversee and train local men in basic construction skills.

CAT's Cairns office was established in 1993 (with one permanent staff member, Mark Moran, increased to three by 1996), following a request from ATSIC's Peninsula Regional Council. However the branch office struggled to establish itself, as Moran said, 'trying to do too much with no money'.[54] In the following year CAT's board and senior management reviewed the viability of its operations in Cairns. The board held its March 1995 meeting in Cairns to assess the situation first hand. As Moran acknowledged to them, 'We have been dealing with rolling back to back consultancies including overseas work. This does not leave time to reflect and consolidate'.[55] At a later board meeting in Cairns in 1997, Moran admitted CAT's ongoing failure to obtain either ATSIC or Queensland state government funding, and explained that the Cairns office still 'operates basically on a fee for service'.[56] Moran and his local staff believed that in this they were doing ground-breaking work, as their community planning projects attested. However, when board members insisted that Cairns justify its existence, and complained that staff in Alice Springs had little idea what the regional office did, Cairns staffers reacted indignantly and emphasised that CAT Inc needed to let them know what head office had in the pipeline, 'otherwise we don't know ... the full range of activity that we're involved in'.[57] Moran left CAT in 1997, disappointed, though he would later return. The tensions that triggered Moran's departure were still evident at the end of the decade as CAT's board sought to recruit an ATSIC representative from north Queensland in order to forge more effective links with the region.

Whereas community design work underpinned the early operations of the Cairns office, homelands maintenance work – especially on roads – was responsible for the opening of CAT's office at Derby in Western Australia. Bray and Kroker visited the homelands around Broome and Derby in May 1998, and Walker made a follow-up visit later in the year. The visitors encountered widespread dissatisfaction with the way in which state and federal funding for services was being implemented. As ATSIC's Derby chairman said to Bray,

> These bloody building contractors push up their prices when it's for an Aboriginal Community. These people rip off the Government and Aboriginal people suffer and get the blame for waste of money. The Resource Agencies show no responsibility, they are just not caring, they dole it out, no responsibility, just a waste of money.[58]

However, it was not clear to CAT how it could attract sufficient funding to do a better job. In a leap of faith, the Derby office (initially operating from one room) was established by trusted CAT staffer Keith Bowden, who set about finding work and funding. He quickly generated almost $1 million in civil works requests from local ATSIC officials while driving thousands of kilometres throughout the Kimberleys, plotting GPS coordinates in order to generate the first reliable map of the strung-out settlements and tracks in the region. His hard work would reap dividends in the following decade.

The precarious relationship between Aboriginal aspirations and government funding formulas and delivery mechanisms, and the unfortunate consequences of this for Aboriginal community development organisations such as CAT, were highlighted at Walungurru during the late 1990s. As Walker noted, 'Kintore has been a very much used community by researchers, government and so on. Lots of publicity, lots of short lived large grants, lots of pressure to act and expand [but] with no real participation time and [ongoing resources] available to support processes'.[59] Walungurru, once the ultimate homelands success story in the eyes of policy makers and media analysts, was by the late 1990s being equated with dysfunctional local government and settlement practices, family breakdown and cross-generational tension, alcohol abuse and petrol sniffing. It was increasingly regarded as a 'slum in the desert'.[60] In contrast to other homelands, Walungurru's outstations were losing population, the result in part of difficulties in transportation, and hence poor access to health and education services, and also because there were few organised activities for young men and women. Women especially felt overlooked, and complained that the community council was not providing strong leadership.[61] Local elder, Lindsay Corby, commented in 1997 that there were 'too much of drunken people and sniffers also young kids running amok and causing … damage, and it has to stop'. Cameron George Yuendemu 'said his job as Night Patrol [officer] … has got worse. He works every night and there's always young people causing trouble'.[62]

ATSIC's dissatisfaction with the standard of service delivery at Walungurru led it in 1997 to suspend funding for outstation services, and to appoint CAT as manager to undertake design work, service provision and maintenance for outstation communities and to reform WCC procedures. Matthew Parnell (who had joined CAT's NTRC in the mid 1990s) was appointed project manager, and Keith Bowden became acting council clerk until he was made CAT's Derby manager the following year. They set about overhauling council operations, staffing, financial accounting and service provision to outstations. This was an important test case for CAT's changing operations on the homelands, and CAT managers were well aware that it had been said 'if it works at Kintore, it will work anywhere'.[63] However, the relationship soured late in 1998 when, after local elections, the new community council and its council clerk voiced criticisms that were 'damaging to CAT's relationship with the community and … made [CAT's] role there rather untenable'.[64] The council clerk (at that time an employee of CAT based in the community) accused CAT of 'not living up to the expectations of the WCC and the community'.[65] Bray led a CAT delegation to meet the community council and 'ask do they still want to work with CAT'.[66] Nevertheless, CAT's board decided early in 1999 to cease operations in Walungurru. It was an unhappy end to almost 20 years of community development work in the region.

* * * * * * * * * * * * * * *

CAT admitted in 1993, early in its struggle to develop and maintain an effective role in places like Walungurru without the necessary funds, that its emerging role of national leadership in Aboriginal community development was being sustained 'by overextending resources [and] by undertaking project management consultancies in order to draw income sufficient to employ additional staff and carry out research'. This make-do strategy, CAT acknowledged, characterised the way in which 'Aboriginal people have generally organised themselves into service provision agencies and resource centres (particularly in support of outstation developments), [but this could not generate] the resources to back up commissioned research and turn it into outcomes that happen'. Thus in terms both of investment and return, CAT's strategy was deemed 'unviable' in the longer term.[67] Ongoing federal funding was needed to underpin the national scope of CAT's homelands activities, and to enable it 'to develop a more comprehensive understanding of the relationships between appropriate technology and the empowerment of indigenous and remote communities'.[68]

ATSIC's funding agreement with CAT in 1994 to establish the NTRC was intended to address these crucial twin needs. Both partners intended that the million-dollar NTRC would function as a national information clearinghouse and research hub, and would work with Aboriginal communities to provide appropriate technology solutions to problems with water and energy supply, sanitation, waste disposal, shelter, food preparation, transport and communication. As CAT explained through its new *Our Place* magazine, the 'main goal of the NTRC is to empower Aboriginal and Torres Strait Islander people to control and use technologies in ways that improve health and housing, and increase training and employment opportunities'.[69] What was not said in public was perhaps equally important: there were no precedents in Australia for such an entity as the NTRC in Aboriginal community development. Thus, as Seemann, the NTRC's initial acting manager realised, CAT's bold experiment 'was unique' in Australia.[70]

By establishing the NTRC, CAT embraced 'the need to rethink the traditional focus of CAT' in changing times. As Porter recognised,

> CAT's traditional identity rests on its demonstrated competency to design and manufacture products that are appropriate and have a lasting impact on the lives of communities. Yet the needs of remote Indigenous communities have changed considerably, partly as a result of increased awareness of and access to technologies available to the rest of Australia. Many of CAT's products were no longer desired.[71]

In launching the NTRC, CAT also 'rejected a "technical expert model" for the organisation, whereby it would become merely another research and consultancy agency competing in the general marketplace',[72] necessarily limited in aspiration to achieving the outcomes specified by each tender. This conventional approach to solving Aboriginal 'problems' tended not to engage deeply with the communities under scrutiny, and to generate a one-way flow of expert technical knowledge that had little relevance to Aboriginal people. The NTRC sought to overcome this impasse in part by channelling more useful information into communities. Issues of *Our Place* were therefore filled with details about forthcoming seminars and workshops, practical information about renewable technologies, water and fire management, outdoor cooking and kitchen design, dust control, cooling methods and maintenance of septic tanks, batteries and outboard motors, as well as stories about successful community development activities. However, CAT went an important step further and sought also to consolidate its 'role as interpreter and advocate for the technical service needs of remote Aboriginal and Torres Strait Islander communities', by developing a two-way flow of knowledge that would 'empower Aboriginal and Torres Strait Islander people to control and use technologies that contribute to sustainable health, housing, employment and technical training outcomes'.[73] As *Our Place* magazine noted, 'There was a lot of research being done but there was not a lot happening at the community end. In some cases, the work that was done didn't reflect the way Indigenous people wanted to approach the problem'.[74]

ATSIC and CAT realised that for their goals to be fulfilled, the NTRC would also need to address the difficult policy environment and implementation framework within which Aboriginal and Torres Strait Islander organisations operated. As CAT had grown, so paradoxically the limits of its influence upon government and the negative impacts of shifting government thinking upon CAT's strategic direction had become ever more apparent. At a workshop with NTRC staff, Porter suggested that the centre should act as 'a filter' between CAT and a broad network of communities, organisations, and individuals in order to fashion a clearer voice on behalf of Aboriginal Australians and to represent them more forcefully to government.[75] This would enable 'Aboriginal and Torres Strait Islander people to take greater control over the technology, livelihood and welfare options available to them'.[76] Taylor's boss at ATSIC, Colin Plowman, who had helped to engineer the NTRC's establishment and was a member of its national advisory committee, urged that the centre's *Our Place* magazine should be used 'to communicate with Indigenous communities, and give evidence that the NTRC is trying to focus on remote Australia as well as Alice Springs'. However, Plowman added,

> More important ... is the public recognition from politicians. He [Plowman] believed that something like 'Our Place' can show them that something is happening and they will then be willing to support it ... To sum up, he believed that one of the key things is general public recognition, including from politicians.[77]

The NTRC had one other purpose: to channel CAT's deepening knowledge of international thinking in appropriate technology and community development, especially in Sub-Saharan Africa and Southeast Asia, into Australian research and policy debate. Porter reminded CAT of this expectation during his review, and urged CAT to draw upon its links overseas, where 'organisations with a traditional role as "tester and disseminator" of reliable, appropriate technologies were increasingly seeing their role within a broader context of community development, both in terms of policy and practice'.[78]

Seemann relished such overseas input: he accumulated data about barefoot surgeons in Botswana, community health workers in China, community veterinarians in Bangladesh, forestry workers in Kenya and water-testing projects in Indian villages. However, as CAT's ATWORK manager, Ron Talbot, noted in response to Porter's review, most CAT staff were more insular:

> Some of us may suffer from the inward looking view that Australian Indigenous community development problems are unique and possibly "too difficult" to solve. The knowledge that there are others out there dealing with similar situations is valuable especially if we can make contact, comparisons etc.[79]

CAT's initial response to Porter's recommendations concentrated on issues of immediate concern that were inconsequential in the longer term, such as the need to drop the word 'centre' so as to remove the confusion of having one national centre located within another. However, exciting developments were already happening regardless of name at the NTRC. Technologically savvy, yet plainly expressed information was being channelled into remote communities. An influx of new and highly trained staff was recruited to CAT, and notwithstanding some initial confusion that 'we didn't quite know what we were supposed to be doing',[80] their presence in the NTRC considerably strengthened CAT's research capabilities. High-quality, applied research was nurtured by Bob Lloyd, the NTRC's manager of research design and planning, who worked for CAT from 1996 until the end of the decade when he moved to Murdoch University. The NTRC undertook cutting-edge studies on house design, shower blocks and hot water facilities, outdoor stoves and cooking aids, water treatment, washing machines, toilets and sewage treatment. As Seemann pointed out, this was really hard-grunt work: by persisting in it CAT 'drove innovation based in research, not based on established standards'. They sometimes 'faced the wall' for not conforming to those standards, but CAT countered that its research 'suggested the standards are not good enough'.[81] The outcomes of this new research, especially in the field of renewable energy, would become more clearly apparent in the following decade. These early technical evaluations complemented the work of Health Habitat in the north-west of South Australia as it developed its Housing for Health initiative,[82] and CAT's research outcomes were referenced extensively in the first National Indigenous Housing Design Guide.

In the meantime, however, research initiatives driven by the NTRC reinforced CAT's credibility as a national hub for appropriate technology, and thus its ability to build partnerships with high-profile organisations around

Australia. Collaborative links were established with research groups at RMIT in Melbourne, the University of Adelaide, Murdoch University and Curtin University in Perth, and the Cooperative Research Centre for Water Quality and Treatment (established in 1995). The most important of these collaborations was CAT's role as a partner organisation in the successful funding application through the Australian government's Cooperative Research Centre programme to establish ACRE in 1997. ACRE's objectives closely matched those of the NTRC and ATSIC, and as a result Taylor increasingly used his influence at ATSIC to facilitate this developing partnership in renewable energy research. In 1999, for example, CAT's board passed a resolution of thanks to Taylor for funding CAT to hold a national forum on energy systems for remote communities.[83] The results of that forum, and of Walker and Taylor's subsequent delicate negotiations with the federal Cabinet and the Australian Democrats, who held the balance of power in the Australian Senate, in the lead-up to the introduction of the GST in 1999, found expression in a $16 million funding agreement over four years from 2002 to direct diesel excise into an ATSIC–ACRE–CAT programme for better renewable energy resourcing for remote communities. This programme, translated into Bushlight, would transform CAT's operations in the next decade.

The NTRC's applied research activities also prompted a phone call from CSIRO's representative on the NTRC's national advisory committee in 1996 to Paul Wand, vice-president and head of Aboriginal relations in the new mining conglomerate RTZ–CRA. The company had formed in 1995 when the international Rio Tinto–Zinc Corporation merged with its subsidiary Conzinc Riotinto of Australia. The new company was renamed Rio Tinto in 1997. Its CEO, Leon Davis, was determined to win market share by distinguishing the company in progressive ways from the mainstream mining industry in Australia. Sensitivity to the viewpoints of Aboriginal Australians, and endorsement of the Mabo and Wik judgements on native title, could be exploited as clear points of difference from their mining competitors. Wand therefore responded readily to the suggestion that he contact CAT, and a meeting was arranged at Melbourne airport between him and Walker. Wand followed up with a visit to Alice Springs. RTZ–CRA recognised that CAT had credibility: it had an Aboriginal board of management, was 'very well networked in the Australian Aboriginal community', and its NTRC national advisory committee comprised national leaders from the professions, the higher education sector and government.[84] Wand agreed to provide leadership training to CAT in order to improve its management practices, and also to extend the company's sponsorship of Aboriginal community development activities (funding for which he judged in collaboration with ATSIC chairperson Lowitja O'Donoghue) by creating a fellowship programme which allowed RTZ–CRA scientists to be seconded to work on CAT's appropriate technology projects. CAT agreed to host visits to Alice Springs and Walungurru by RTZ–CRA executives during cultural awareness training.

The annual Rio Tinto Fellowships were an immediate success, and generated useful fieldwork on dust abatement, rubbish disposal and septic tanks. Wand and Walker also introduced an annual weeklong science programme for Aboriginal children, who were brought to Alice Springs from regions throughout Australia in which Rio Tinto worked. Designated the Akultja Youth Science and Technology Event, the first of these was held in 1999. It was backed by the federal Department of Education, Training and Youth Affairs. Wand suggested early in 1999 that this collaboration be formalised by drawing up a memorandum of understanding. He was approaching retirement and wanted to ensure that the joint ventures continued into the future. CAT's board had never before entered into a formal partnership with the private sector, and was cautious about aligning themselves more closely with such a powerful resource company. They felt that 'Rio Tinto as a large organisation should be aware that CAT is a small organisation and does not want to be overwhelmed or directed by the Company'.[85] Bray, deputy chairman Frank Curtis, and Kroker visited Lowitja O'Donoghue, who had become a Trustee of the Rio Tinto Aboriginal Foundation, to seek her advice. Reassured, the board embarked upon negotiating the terms and wording of the MOU with Rio Tinto, and it was formally signed in Parliament House in Canberra on 30 November 1999. It announced 'a shared commitment to furthering the social and economic development of indigenous people by increasing the awareness, knowledge and application of science and technology', and by harnessing these to improve health and housing options and to capitalise on business, employment and development opportunities for Aboriginal communities.[86]

There was one other important consequence of establishing the NTRC. ATSIC and CAT agreed that the NTRC should work to achieve better outcomes for women and young people, who, as both organisations conceded, had been hitherto marginalised in community development activities. At the urging of the women's unit within ATSIC, the NTRC was given the task of establishing an Aboriginal women's technology network. In mid 1995 two new NTRC staffers, Sonja Peter and Su Groome, won support from CAT's board to implement this agreement by appointing an Aboriginal woman as a full-time coordinator to increase women's participation in decision making about technology and Aboriginal wellbeing, promote women's involvement in CAT and establish an Aboriginal & Torres Strait Islander Women's Technology Network (WTN).

Pending this appointment, Robyn Ellis, a newly appointed ATWORK lecturer in design and technology at CAT, organised a meeting at CAT in October 1995 to establish the beginnings of such a network. Fifteen women representing a variety of organisations and communities attended. The network was formally launched early in 1996, its goal being 'to assist women to take more control over community development and to be more involved in making decisions'.[87] Kroker was recruited to CAT as WTN coordinator. She was an Eastern Arrente woman from Alice Springs who moved to CAT after working for 20 years in higher education administration in Wagga Wagga, New South Wales. She and Ellis worked well together. They were strongly motivated, and pointed out that most Outback technologies remained inappropriate because women had not been included in the design process. They contended that although women historically had organised bushcamp life, since European colonisation they had been 'locked out of what really was traditionally their area'.[88]

Kroker and Ellis set about establishing a national WTN and organising a 'hands-on' national conference, using the slogan 'Is Technology Women's Business?'[89] A series of workshops were held around Australia in the first six months of 1996, from Broome in the west to Port Augusta in the south, Wagga Wagga in the east, and the Torres Strait and Thursday Island in the north, to identify key issues for discussion at the forthcoming conference. A key finding to emerge from these workshops was the widespread concern among women 'that cultural and spiritual elements were often overlooked when designing and developing communities'.[90] The NTRC hosted the first ever Aboriginal and Torres Strait Islander Women's Technology Network Conference in Alice Springs over four days in August 1996. The conference aimed to 'promote the recognition of Indigenous women's knowledge and experience with traditional technology', to generate broader discussion about technologies appropriate to Aboriginal wellbeing, and to formally launch the WTN.[91] Over 500 women took part, representing a wide variety of land and ATSIC regional councils, women's resource centres and organisations, CDEP programmes, government departments and the VET and higher education sectors.

Notwithstanding the women's energy and enthusiasm at the conference, it was inauspicious timing for such a major initiative. In 1996 the federal government cut back funding to ATSIC and other federal agencies that supported Aboriginal women's initiatives. One immediate result was the closure of the ATSIC women's unit that had championed the WTN. Kroker reported to the NTRC national advisory committee in 1997 that the WTN 'has been really slow since the conference'.[92] During the year the WTN established a two-day intensive course for women on domestic technology training, available through

CAT's ATWORK programme and delivered to local communities using CDEP funding. Kroker participated in a number of women's forums and workshops on technology in remote communities during the late 1990s, but she had to juggle this with her new responsibilities as CAT's executive officer. The overall results from the WTN conference were disappointing. In 1998 the NTRC's national advisory committee suggested that CAT survey women's attitudes to technology, and with ATSIC support a national survey was conducted. The results, released in 2000, were sobering: a majority of respondents expressed deep concern about gender exclusion, and a quarter identified a pressing need for women's technical training and for education in practical skills.[93] However, in an increasingly unsympathetic policy environment, women's initiatives faltered and the WTN ceased to function.

****

During the 1990s CAT, in addition to maturing its original research and design activities, embarked upon a new role as a national innovator in technical education for Aboriginal people living in Outback communities. CAT had run a small training programme in automotive mechanics throughout the 1980s, and had begun training courses in outstation communities in conjunction with the installation of products from the Priest Street workshops. These early experiences gelled with Walker's knowledge of community-based service delivery and training in Africa, China, Indonesia and Latin America, and of the Aboriginal Health Worker programme that the Northern Territory Department of Health had developed since 1979. CAT drew upon these precedents and its own training experiences when in the mid 1980s it proposed an Aboriginal Technical Worker (ATWORK) training programme. Walker explained that ATWORK drew upon the '"Barefoot Engineer" approach [and] the "Barefoot Doctor" or "Community Health Worker" strategy now being used world-wide in health care systems'.[94]

Commonwealth funding enabled ATWORK to begin with an initial feasibility study (led by Seemann, who was appointed as ATWORK's project officer in 1987) in 34 communities across three states during 1987–88, which resulted in a draft report in 1989 and a final report in 1990. This led in turn to a multi-million dollar grant under the Commonwealth's Aboriginal Education Program in 1990 to implement a nationally accredited ATWORK programme, developing curriculum resources from Northern Territory Aboriginal communities and delivering a pilot trainee course at Walungurru during 1990–91. Additional support was received from the Northern Territory Employment and Training Authority. In 1992 ATWORK was built into the national TAFE curriculum and was showcased at a national ATWORK conference in Alice Springs in August 1992. ATWORK was the first national technical vocational curriculum to address Aboriginal community needs. Seemann reminisces, 'We won a National Innovation in Curriculum award for the way we combined traditional with contemporary technical knowledge while overcoming a major bias by the trade unions' in favour of mainstream trade courses.[95] In 1994 the first ATWORK graduates received their national certificates in Applied Design and Technology.

The most obvious signs of ATWORK's rapid development during the early 1990s were seen at Priest Street. Although orders for homelands products slowed, expanded workshop facilities were needed for ATWORK trainees and teaching staff. A new ATWORK training workshop was built, and an open day was held in 1994 to show off the new facilities. A demountable was added as ATWORK classes grew. The residential complex at Priest Street was also heavily in demand, with students coming from the Northern Territory, South Australia, Western Australia and Queensland. Bray, its live-in manager, reported that in the six months from June to December 1997, for example, 423 residents had passed through Priest Street. It was at times a tumultuous place. Bray added that because drunks often tried to climb the fence of the residential complex during the night, the fence had been increased in height, barbed wire run along its top and security lights installed.[96]

The most important element in ATWORK's programme, however, was community-based training. On-campus instruction at Priest Street was intended to complement ongoing training within participating communities, so that trainees progressed through a sequence of 'project driven modules, utilising actual problems, issues and developments taking place in a student's community or organisation [and] thus contributing to community development'.[97] Training focused upon the repair and maintenance of buildings and equipment such as lawn mowers, septic tanks, generators, water drains, pumps, radios, solar hot water units, power tools and boats.

ATWORK's conceptual framework rested on three elements: Walker's training in ekistics, the study of viable patterns of human settlements (shared by Seemann, who completed a PhD at the University of New South Wales with Walker as his field supervisor); the exposure of new CAT staff to the canons of appropriate technology and community development (Seemann had studied Schumacher, Gandhi and Lewis Mumford, and Ellis was 'smitten' when she first read Schumacher);[98] and, lastly and most importantly, CAT's first-hand experiences working in Outback communities. ATWORK was thus built upon recognition that the 'tendency in small communities of less than approximately

250 persons (particularly in homelands and outstations) is that individuals tend to require and apply broad skills and tools across community functions', and that these 'skills, technical resources and valued cultural practices [should be] taught in harmony with the existing skills and functions of the community'.[99]

In elaborating this framework Walker drew largely upon the expertise of Seemann, who was ATWORK's project officer until becoming manager of Research and Development for the NTRC in 1995, and Talbot, who was appointed CAT's Manager of Education and Training in 1992. Talbot had initially trained as an electrician at the Wollongong steelworks in New South Wales, and later qualified as a technical teacher at the University of South Australia before becoming heavily involved in Aboriginal education and training in the Northern Territory. He was seconded to CAT from the Territory's Department of Education in 1992 and remained there for 15 years. Talbot helped Seemann to write the ATWORK national curriculum. He also recruited Ellis to start the Women in ATWORK programme.

Seemann and Talbot built into the ATWORK curriculum the concept of 'technacy education' as a parallel strand to mainstream education in literacy and numeracy. Its focus was teaching 'an understanding of the dynamic interacting relationships between people, technology and their environment'.[100] In an Outback context, technacy was therefore necessarily holistic and cross-cultural. Seemann especially delved into both its intellectual underpinnings and practical applications, and inserted the term into the *Macquarie Dictionary*.[101] He described technacy as building into the technical education curriculum 'the training which relates your lifestyle, your goals and your values to the needs that you can identify around you'.[102] This entailed that courses be 'based on a holistic approach to perceiving, teaching, practising and learning technologies in any culture'.[103] Seemann and Talbot required that all course modules be linked to local areas of need that had been clearly identified by participating communities, so that 'students draw their learning resources from the actual issues, problems and developments taking place in their own communities'. They also sought to ensure that course delivery placed 'emphasis on the integration of a variety of technologies, materials and cultural knowledge in order to produce appropriate responses that actually support community functions and cultural activities'.[104]

Translating these ATWORK concepts into practice in remote locations, Ellis set out to demonstrate that 'ATWORK is about applied design … my job was to teach a broad range of technical skills, so there'd be a bit of carpentry, welding, plumbing, concreting, whatever is needed, but specialising in teaching problem solving'.[105] ATWORK's 4WD light trucks were so equipped that staff could set them up 'right next to the bore that needed to be repaired, or the house window that needed to be repaired, or the door that needed to be repaired; we could drive right up to it and work'.[106] ATWORK thus used technical education to address 'living issues', such as how to design door locks that didn't jam because of dust, or rubbish management, or erosion control. At Yeundumu, for example, located to the far north-west of Alice Springs on the remote Tanami Track, Ellis spoke to the local women about ATWORK and explained to them 'that what we had to offer was about grappling with this whitefella technology that keeps breaking, because they all acknowledged ... that the toilet's blocked, and this is broken, the houses are trashed'. Ellis and the women agreed to run an ATWORK training module in which they would produce a set of murals on the walls of the community toilets, showing how they worked, how they should be maintained, and how they could be blocked by inappropriate use. Ellis laughs that 'a mural solved a technical problem', because the toilets did not block again for years.[107] Ellis also regularly offered ATWORK training at Walungurru, instructing women as they renovated houses and collected rubbish. She continued to work in the community even after the breakdown in relations between CAT and the WCC, running an ATWORK training programme to help local women build a museum for women's sacred artefacts.

Such ATWORK activities were underpinned by the proposition that 'training in technical skills, concepts and capacities relevant to living in remote communities is as important to Aboriginal people as literacy and numeracy, and fundamental to community self sufficiency and self determination'.[108] Talbot was adamant that ATWORK courses must help students with 'articulating their Aboriginality', and that to this end training had to adhere to a 'both-ways' educational approach that combined 'culture identity and non-Aboriginal influences'.[109] CAT argued that these outcomes were hampered by conventional technical training, because formal trade-based subjects did 'not support the lifestyle of people in remote communities ... We estimate that somewhere between 90 and 95 per cent of the people who live in remote communities are totally unserviced by existing programs'.[110]

As Seemann contended, a one-size-fits-all approach to technical training throughout Australia was insufficient: 'simply providing the technical skills and a few tools is not enough', he said; instead, Aboriginal communities needed 'Curricula founded on the precise needs recurring at the community level'.[111] Those needs had been pinpointed during CAT's ATWORK feasibility study fieldwork in 1987, which found that:

> The western concept of a certified tradesperson is totally inadequate for remote communities. It is impossible to organise the necessary training and

> there is usually no job available for such a person anyway. However, there are quite a few smaller technical jobs performed by community members (minor construction, car care, plumbing, sewing) and quite a few more performed by outsiders ... on contract basis that could be performed by community members with a little training.[112]

In Yirrkala in East Arnhem Land, for example, it was reported that 'Formal qualifications are not really vital in an Aboriginal community. Much of the work can be done without a tradesperson [if basic general technical skills were put in place]'.[113] Pointing to these findings, Walker concluded:

> When you look at the most used pieces of technology in the domestic spaces and you try to equate that with the sort of courses that are available through TAFE or a whole range of other formal or informal providers, you will find that there is just no relationship between what people are doing in their lives and the courses and the training programs that are available.[114]

Notwithstanding the clear need for ATWORK training in Outback communities, and the 'humming' mood of enthusiasm amongst ATWORK instructors,[115] it became increasingly evident during the mid 1990s that take up of the new programme was hampered by three problems. Firstly, although formal qualifications were not needed within homelands communities, formal education pathways into wider employment markets were clearly necessary if Aboriginal communities were to break their dependence on welfare. By the middle of the decade CAT was acknowledging the need to develop articulation and credit transfer agreements with other TAFE providers and universities, to enable students to move from ATWORK into higher-skill education programmes. CAT also recognised that attention needed to be paid to incorporating English and Aboriginal language competencies into ATWORK training and to maximising participation by Aboriginal women in the programme.

Secondly, there were differences of opinion even within CAT about the ATWORK programme. On the one hand Talbot worried that CAT might compromise ATWORK's goal of Aboriginal empowerment in order to sustain its funding stream from government:

> CAT training has to be careful about the 'how' and 'why' in its response to community training needs. I see a danger with CAT falling into the trap of being driven by such things as Government Competitive Tendering for training (particularly in terms of picking up more dollars as opposed to community development needs under CAT's vision and purpose), training for training's

> sake, poor interpretation of CAT's vision and purpose by E&T staff and clients, a mainstream view of technical training by CAT's E&T staff and clients.[116]

On the other hand, one sceptical inhouse reviewer queried the intellectual basis of ATWORK, pointing out that half the Aboriginal staff in the programme did not fully understand what the core concept of 'technacy' meant.[117] Another ATWORK trainer contended that, more generally, none of the teaching staff knew 'what appropriate technology means', and he pointed to a 'big wall' between the training staff at Priest Street and the other CAT staff.[118] The space at Priest Street between the trainers' facilities and CAT's administration was known idiomatically as the 'Gaza strip'.[119] During the initial feasibility study for ATWORK another reviewer expressed concern that although the proposed

> vocational type of training ... will be of great assistance to communities, national standards and certification should be the ultimate goal ... Aboriginal people have long been made to feel like "second class" citizens. I would not like to see them become "second class" tradesmen and technicians as well. This is the inherent danger of the [ATWORK] approach.[120]

Thirdly, by far the greatest impediment to the long-term success of ATWORK was changing federal government policy. After Keating became prime minister in December 1991, policy emphasis increasingly focused upon Australian economic competitiveness, and to this end upon mainstream workforce training and participation. More and more, the expectation was that TAFE training should produce the specialist trade qualifications that were in demand in the mainstream economy, and this impacted badly on funding to develop generalist competencies in Outback Australia. By 1996 CAT was wrestling with the problem of maintaining government funding for ATWORK and the related problem of stemming declining student enrolments. The board struggled to articulate the difference and value of ATWORK compared with 'mainstream' approaches to technical training.[121] The election of Howard as prime minister in March 1996 intensified these pressures. It was not that the new federal government was unsympathetic to CAT's activities: the first Minister for Aboriginal Affairs to visit Priest Street was John Herron, a member of the Howard government, who visited in 1997, and David Kemp, Howard's Minister for Employment, Education, Training and Youth Affairs, visited CAT's ATWORK facilities in 1996. However, the Howard government's emphasis upon trade-based technical training forced Talbot reluctantly into watering down successive versions of the ATWORK national curriculum until only a 'smattering' of technacy remained.[122]

1. If you want to be someone and do something just give yourself a name and place a sign on the door. This sign in 1981 signalled the arrival of the Centre for Appropriate Technology well before it was formally acknowledged.

2. The owners of this vehicle have a problem with the gearbox which is now removed and is sitting on the rug. Imagine you are a highly qualified mechanic. What skills and knowledge would you be able to offer the Aboriginal owner? What we know and value is heavily dependent on context. Remote Australia constantly challenges our thinking and understanding of technology.

3. Jim Bray and Bruce Walker at Priest Street just prior to CAT's move to its new home at the Desert Peoples Centre on the Desert Knowledge Precinct.

4. *Almost 20 years of community development work at Walungurru and the Kintore Outstations.*

Left: An early dwelling at Ininti Outstation near Kintore.

Below centre: The use of an improvised jib crane to raise roof beams for the portal frame housing built on the Kintore Outstations.
Bottom: The Kintore building team constructed all of the portal frame houses for the Kintore Outstations.

5. Completed portal frame house at Turkey Tolson's outstation.

6. The Western Desert Art movement flourished with the creation of the eight outstations that enabled the painters to get back on country. This group is painting at Desert Bore Outstation.

7. Community meeting adjacent to the six VIP latrines installed at Kintore in April 1984, before the school was built.

8. The same six VIP latrines, Kintore School 2011 — painted and regarded as heritage items by the community.

9. Herbert Bloomfield fitting out one of the original round showers built for the emerging community at Kintore (Walungurru).

10. *CAT constructed a number of innovative sprayed concrete cluster houses in collaboration with Melbourne architect Mario Bernardi.*

The Kintore Housing team assembling the foam and mesh monolite panels before spraying with concrete.

The Kintore Housing team constructing the septic tank for the monolite cluster house.

The completed monolite cluster housing, Kintore.

11. Transporting the handpump for installation at Kunapulla Outstation in the Petermann Ranges near Docker River.

12. The first handpump installed at the newly formed community at Kintore. Pintupi people moved from Papunya to Kintore in early 1981 once this handpump was installed.

13. Robyn Grey-Gardner conducting a workshop at Port Stewart, Qld, using the Community Water Planner Field Guide developed by CAT, Water Quality Research Australia (WQRA) and the National Water Commission.

14. Mark Moran and Mapoon community members discussing plans for peoples re-settlement along the coast at Old Mapoon. Settlement and community planning was an important part of CAT's North Queensland work.

15. First trainees at CAT, 1983 with lecturer John Appleyard.

16. Construction of the Women's Keeping Place by the Kintore Women in ATWORK trainees.

17. CAT residential buildings in Priest Street used by a student group from Santa Teresa.

18. Ron Talbot demonstrating the use of the community modelling sandpit as part of the ATWORK training program.

19. Automotive and mechanical skills students at Priest Street workshop 2006.

20. CAT ATAF ablution with CAT hot water chipheater and hand powered washing machine at Bonya community.

21. Max McKenzie, Matthew Palmer, Trevor Corbett and Bradley Corbett proudly displaying a production run of chipheaters in the CAT Enterprise Training Workshop.

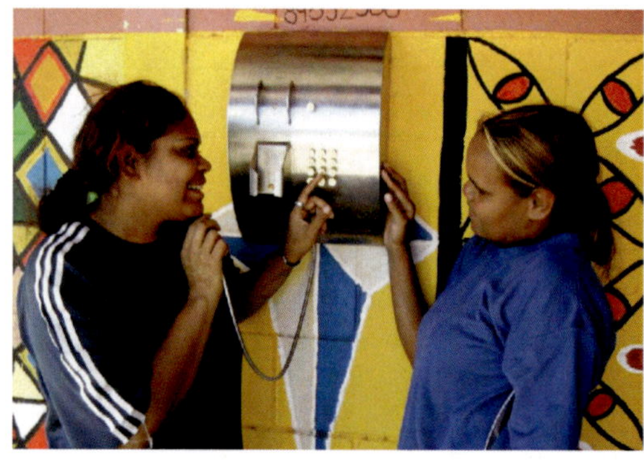

22. The CAT Community Access phone in use in the town camps, Alice Springs. This innovation allowed a standard residential service phone to be installed in a public space outside a house.

23. Mobility Aid in use at Ininti.

24. Tony Renehan demonstrates the operation of the CAT drum oven to HRH Prince Charles during his visit to CAT, 2 March 2005.

25. Kenny Kunoth and Bruce Walker releasing the mould used for mud brick production at the Utopia Homeland –Alparrinya Apungalindum.
Over the years CAT experimented with many building materials and construction techniques.

26. Pole house construction underway at Port Stewart in Far North Queensland.

27. Fixing the laundry sign at Yuendumu as part of the Women in ATWORK program 2002.

28. Delivering a renewable energy solar pack to Jupiter Well through sand dunes west of Kiwirrkurra WA.

29. Dr Bob Lloyd checking a monitoring unit on a solar hot water heater being trialled as part of CAT's hot water research project.

30. Residents at Birany Birany in front of their new solar array, commissioned as part of CAT's Bushlight project.

31. Bushlight System training being conducted by Phyllis Batumbil at Rorrowuy, Arnhem Land NT.

32. Launch of the National Technology Resource Centre Clearinghouse 1995. Colin Plowman from ATSIC and ATSIC commissioners John Pattersen and Charles Perkins, CAT Board members James Bray, Moogie Patu, Peter Ferguson and Bruce Walker (standing).

33. The first CAT/Rio Tinto Akultja Youth event presentation evening with Jenny Kroker, CAT Board Member.

34. Go Kart building and racing was a highlight activity at the Akaltye Youth Event.

35. Aerial view of the Desert Knowledge Precinct Alice Springs.

36. Arrernte men dance at the Opening of the Desert Peoples Centre 2010.

Walker later conceded that 'we were never able to get technacy to stick' because of VET's increasing emphasis upon trade training rather than generic skills training.[123] Seemann left CAT in 1998, initially to take up a post-doctoral fellowship in the CSIRO, and then to teach at Southern Cross University. Talbot persisted in his line of argument at CAT. In a 1998 conference paper entitled 'One Step Forward and Two Steps Back', he and Kroker drew attention to the continuing failure of mainstream technical training to address the problems of poor technological knowledge and deteriorating health and wellbeing in Aboriginal communities, and warned that the federal government's TAFE programmes, delivered through a 'new VET National Training Framework, with its emphasis upon mainstream industry needs and the rationalisation of standards, is hindered by blinkers' when applied to Outback Australia.[124] Walker, shaking his head over VET's evolving framework, declared, 'You don't need that amount of paper to teach someone how to drive a nail!'[125]

3. Core Principles

The decade of the 1990s thus delivered fundamental change and mixed outcomes at CAT. The greatly increasing scale and influence of its operations won respect from the 'bush mob' and was recognised in Canberra, the capital cities and overseas, but they also generated divided opinions within CAT itself about aims and methods, and they did not free the organisation from dependency upon a public policy framework that often seemed geared to frustrate CAT's best endeavours. In 1995 one CAT staffer declared pessimistically that the organisation was characterised by 'low staff morale, and lack of direction and collectively understood vision'.[126] Notwithstanding the altering and challenging circumstances within which CAT operated, a set of core principles – inherited from CAT's early operations during the 1980s, and evolving to meet the changing circumstances of the 1990s – continued to shape CAT's activities. These principles were highlighted by the Human Rights Water Study that Walker worked on through much of the 1990s.

The Central Land Council told the federal Race Discrimination Commissioner, Irene Moss, in 1990 that 'the development of the entire outstation movement depends upon the supply of water'.[127] Water authorities appeared unable or unwilling to provide adequate supplies, or attempted to do so using intrusive and costly technologies. Moss concurred with the Land Council's assessment, having during 1987–88 led an inquiry into living conditions in the Aboriginal community of Toomelah on the New South Wales-Queensland border in the aftermath of the 1987 race riots in the border towns of Goondiwindi, Boggabilla and Toomelah. As she later recalled, 'The image stayed with me of the Aboriginal townspeople queuing up at standpipes to fill receptacles during the short periods of the day when the water was actually turned on'.[128] After releasing the Toomelah report, Moss received complaints from other Aboriginal communities about inadequate water and sanitation. In response, early in 1989 she invited applications to comprehensively study water issues in Aboriginal communities. Moss wanted two questions answered. Firstly, were inadequacies in water supply undermining Aboriginal wellbeing? The study was thus intended to investigate concerns about salinity, levels of bacteria and minerals, unreliable and inappropriate technology, government maladministration and insufficient consultation with communities. Secondly, did inadequacies in water supply breach international covenants on basic human rights to which the Australian government was a signatory? Moss was keenly aware that in the representations to her about poor living conditions after the release of her Toomelah report, the 'most common of these problems concerned something most Australians take for granted as a basic human right: *the right to a decent water supply*'.[129]

The water report contract was awarded to Walker in June 1989, in collaboration with the Institute of Environmental Science at Murdoch University in Western Australia. Walker's referees were the director of the Central Australian Aboriginal Congress, representatives of the Equal Opportunity Commission in South Australia and the Water Authority of Western Australia, and CSIRO scientist Basil Hetzel, who had watched with interest CAT's development out of the TAGAL initiatives that he had led during the late 1970s. Walker won agreement to work intensively with eight Aboriginal communities: Punmu and Coonana in Western Australia, Yalata and Oak Valley in South Australia, Tingha and Dareton in New South Wales, Doomadgee in Queensland, and Mpweringe-Arnapipe in the Northern Territory. In 1991, at Walker's request and with the backing of Gerry Hand, the Minister for Aboriginal Affairs, the study was extended to enable additional fieldwork in two Torres Strait Islander communities: Coconut Island and Boigu Island. With the Commission's backing and additional support from ATSIC, CAT hosted a National Water Forum in 1993. Moss released the Water Study at a ceremony in Sydney and tabled it in the federal parliament in the last days of her term in May 1994. Walker reviewed his main report at the request of the Human Rights and Equal Opportunity Commission in 1999, with the active support of Taylor at ATSIC, to facilitate his work in the ten communities.

When Moss launched the main water report in 1994 she used the example of Yalata in South Australia to highlight the findings and the seven main recommendations of Walker's report. Government experts at Yalata had installed a high-tech desalination plant to process bore water, rather than opting for the simple option of harnessing rainwater runoff. Moss noted that for years, 'the technology of the desalination plant and its constant and expensive repairs have dominated the agenda at Yalata. Indeed, the health of the people became secondary to the technology itself'. Moreover, she added, the 'technology disempowers and alienates the Aboriginal community. The community feel inadequate because they cannot fix the technology when it is broken; they have lost faith in high-tech solutions; and at the end of the day, they are still left with undrinkable water'.[130]

The federal Race Discrimination Commissioner drew particular attention to three recommendations in Walker's report. Firstly, that decisions about water supply and quality 'give effect to self-determination through real community control'; secondly, that government 'standards and procedures [should] not conflict with cultural aspirations and lifestyles'; and thirdly, that government, rather than clinging to a 'blanket interpretation' of the concept of equality of service – 'which is generally taken to mean that the government must provide exactly the same services to all people wherever they are' – should instead 'focus more on the "equity of outcome"', ensuring that all Australians received their right to a guaranteed and adequate water supply, but recognising that the best technology to deliver that right would vary according to location, settlement type and lifestyle.[131]

Moss's comments have a broader currency. Walker's water project, and the many related activities undertaken by his staff at CAT throughout the 1990s, were grounded in three core principles: appropriate technology, Aboriginal empowerment and participatory two-way community development. They are considered in turn in the following sections.

****

Firstly, CAT maintained its advocacy of appropriate technology products and approaches throughout the turbulent decade of the 1990s. As Porter noted in his ATSIC review, 'CAT's identity traditionally rests on a deserved reputation for practical technical research and the manufacture of well designed products that are technically and culturally appropriate to indigenous communities'.[132] When Walker won the Clunies Ross National Science and Technology Award in 1999 for resolving livelihood issues in remote Aboriginal communities, public attention again focused on the pioneer hand pump he had installed at Walungurru in 1981, and the other CAT appropriate technology products that followed.

However, by 1999 CAT's original role as a provider of homelands products had substantially changed. This was the result in part of changing government policies and funding schemes, which provided new opportunities for CAT in research, communication and training (for example, through ATWORK and the NTRC), but which also resulted in private providers supplying CAT-designed products more cheaply than could CAT. It was also the result of continuing frictions within Aboriginal communities over competing technologies. As Porter recognised, even in small communities at the far end of supply chains, expectations were changing as a result of 'improved access to and an aspirational trend toward more sophisticated products in remote communities: air conditioners, instant hot water, flush toilets, etc.'[133]

Walker acknowledged at the beginning of the decade, when asked about the continuing appropriateness of CAT's iconic VIP toilet, 'That is one of the problems. People think they are getting second-class options'.[134] At Walungurru, for example, the school principal Ralph Folds noted that 'Pintupi insist they [VIPs] are not "proper toilets"', and that CAT's '"appropriate" houses [are] the first to be abandoned'. Folds maintained that

> The debris of rejected internationally acclaimed 'appropriate' technology, such as hand-operated mechanical washing machines and wood-fired water heaters, which litter their settlements, is ample testament to failed attempts at providing culturally and environmentally appropriate technology. Despite the inconvenience of such hardware Pintupi are expected to welcome it into their modern lives because it can be controlled and integrated in the same way their traditional technologies were.[135]

Some residents at Walungurru submitted a formal letter of protest in 1991 that 'We people who live at Kintore are very upset at the sort of houses we have ... We look at the houses for the teachers and the Sisters and all the white people who come to live here and they keep getting better houses'. The complainants were especially upset that whitefella houses had air conditioning and internal stoves rather than external CAT bush ovens.[136]

Communities rarely speak in one voice, however, and Seemann responded to the Walungurru complaint by contacting Smithy Zimran, the WCC community clerk and a strong supporter of CAT's activities, to ask his advice 'so that we can better understand the concerns expressed'.[137] Moreover, in other communities there was unambiguous endorsement of appropriate technology concepts. For example, representatives at Galiwin'ku, the largest settlement on Elcho Island, off the eastern coast of Arnhem Land, objected to the use of diesel generators to supply electricity, and called for the introduction of

> smaller, simplified more appropriate technology to the community. Is it appropriate to develop replicas of Darwin suburbs on communities; more complicated things [bring with them] more likelihood of breakdowns, more dependence on outside contractors to fix and less possibility of employment prospects for local residents.[138]

CAT staff were keenly aware of the delicate balancing act that they had constantly to tread between such opinions. As one CAT engineer asked after visiting Walungurru in 1995, 'What supportive role can CAT perform for the community rather than the present one that appears to be one of giving the community what it thinks is best for it?'[139]

CAT staff had already confronted these Outback complexities – mainstream technology failure and divided opinions about appropriate technology alternatives – during the 1980s. During the 1990s, they moved cautiously towards developing a two-way flow of knowledge (evident especially in ATWORK and CAT's community planning initiatives in north Queensland) that could pinpoint the diverse cultural contexts within which disagreements about technological interventions occurred, and build Aboriginal knowledge into the design and implementation of appropriate technologies. As Walker quipped in 1991 while working on the water project, 'The medicinal and spiritual association with water and water sites, their relationship to the seasons and rainmaking etc., are as significant in the minds of many "traditional" people as the NH&MRC Water Quality Guidelines are to non-aboriginal people'.[140]

* * * * * * * * * * * * * * *

Secondly, CAT remained committed to self-determination and empowerment. It regarded appropriate technology – both the products and the consultative and training processes that conditioned their design and use – as necessarily informed by the ongoing dynamics of evolving Aboriginal knowledge and culture, driven by Aboriginal choice, and directed towards improving Aboriginal wellbeing and maximising Aboriginal livelihood options. It followed that appropriate technology should be a tool of Aboriginal self-determination and empowerment. CAT's initial statement of purpose during the early 1990s – 'To research, design, develop, produce and teach technologies appropriate to Aboriginal lifestyle, thereby enhancing self-reliance, self-determination and enterprise of Aboriginal people'[141] – was strengthened in its 1996 Strategic Plan for the years 1997–2000 by a mission statement that CAT's purpose was 'To empower Indigenous people, who live in remote communities, to achieve self determination and enterprise leading to social and economic development.'[142] As Porter recognised, CAT's activities were driven by a desire to develop 'a more comprehensive understanding of the relations between appropriate technology and the empowerment of indigenous, remote communities'.[143]

The shifting emphasis from 'self determination' to 'empowerment' had less to do with abstract concepts from academic and political debate than with the first-hand experience of attempting to translate Aboriginal choice into improved wellbeing and sustainable livelihoods. This was reinforced when Bray took a study tour in Canada during 1997, examining indigenous health, housing and education, and focusing especially upon youth substance abuse and suicide. Bray, and also Moran who undertook a Churchill Fellowship in Canada, thought that 'the idea of wellness' that indigenous organisations in Canada pursued was particularly appropriate in Outback Australia.[144] By contrast, CAT's Aboriginal board had little patience for well-intentioned whitefella platitudes. During one planning meeting during which the need for 'cultural appropriateness' was raised, Noel Hayes

> jumped on the table and thumped it and said 'you can't tell me what's culturally appropriate, that's for me to decide, so stay away from it! We don't want to hear this word what's culturally appropriate or whatever. You can't take our culture [till] we give up our culture. If we lose it, we lose it; you can't take it'.[145]

CAT's approach was also informed by Walker's fieldwork across Australia during the 1990s when he was working on the water project. Walker found widespread inequalities of access, ongoing discrimination and what he called latent racism 'institutionalised in the value system of technocrats charged with responsibility for water supply'.[146] CAT ran into similar mindsets in other technology areas, as its focus on alternative design, community participation and use and training of local labour through ATWORK, came under pressure as a result of shifting government policies and use of contract labour. House design, construction, maintenance and energy supply increasingly relied on outside interventions that mirrored mainstream expectations and practices. At Walungurru, for example, Clive Scollay (who was appointed manager of the NTRC – renamed the National Technical Services Clearinghouse – in 1998) expressed concern in 1990 that because of imported design principles 'Kintore appears to have a rather rigid layout based principally upon a "suburban" model planned around proximity to services rather than to any careful analysis of community lifestyle'.[147]

Walker contended that such policies amounted to a renewed form of assimilationism that was based on the assumption 'that Aborigines must ultimately enter the main stream of Australian life [and] behave more like Europeans and as a consequence ... relinquish much of the lifestyle which they still value highly. The pressure to move in this direction is fuelled by technology'. The measures of progress used by government agencies, he said, reflected 'a western view of development. Such measures of development impose many anxieties and frustrations on Aboriginal people when they do not effectively contribute to their development. A house is one of the most visible of these symbols'.[148] Walker argued strongly in his water report that policy for Aboriginal Australians should be rewritten based upon the 'principle of self-determination [which] raises the profile from consultation and negotiation with communities from being an optional and minor part of the process to being the fundamental starting point and lynch pin for entire programs'.[149]

However, CAT's work in communities demonstrated that well-intentioned talk about Aboriginal self-determination and empowerment, finding expression in technological interventions that sought to deliver citizenship entitlements to all Australians, could paradoxically entrench Aboriginal dependence and undermine empowerment. CAT had no issue with 'ATSIC's position ... that adequate housing with water, sanitation and power is a citizenship right'.[150] However, Walker advocated a nuanced appreciation of human rights that upheld equity principles while allowing for diversity of choice and outcome. He cautioned against interpreting rights to mean universal access to identical technologies and services irrespective of geographical location. As he told a Senate committee in 1991:

> It should not be just assumed that by taking the systems that work in Sydney, Melbourne, Canberra and all the other cities and planting them in remote communities that they will improve things. There are differences that occur not because people are Aboriginal but purely because they are remote ... To try to replicate in the bush what you have in the city is in many ways an unequal exercise.[151]

Walker worried that such approaches not only disregarded Aboriginal culture but undermined the opportunities for Aboriginal self-determination by increasing dependence upon outside service providers and technicians. He argued in his water report that

> misunderstanding of the intent of race discrimination legislation and human rights provisions have led some service providers and receivers to the view that all people should receive the same services, irrespective of their race. This

> outlook frequently results in technologies and programs being introduced to Aboriginal and Torres Strait Islander communities to achieve lifestyle changes and developments seen as desirable by non-indigenous people [but which] may well work against the achievement of goals of equity.[152]

As Tim Rowse would write in 2002, 'equity' paradigms provided a limiting and inappropriate policy framework for appreciating 'cultural difference' as rather 'a chosen destiny than a misfortune'.[153]

****

Thirdly, CAT remained committed to participatory community development; its evolving thinking about appropriate technology design and Aboriginal empowerment during the 1990s greatly matured its approach to community development. Porter noted that

> One of the key strengths of CAT is its ability to utilise local and indigenous knowledge in its services. The evaluation at Mapoon showed clear evidence in the planning process of appropriate consultation, integrated use of local knowledge and history, gender equity and involvement of youth and children in the planning process.[154]

In Walker's opinion, CAT's reputation in the community development field was first established during the mid 1990s through its participatory community planning work at Mapoon and other locations in northern Queensland.[155]

Mapoon is some 80 kilometres north of Weipa, on the traditional lands of the Tjungundji people on the western coast of Cape York peninsula. An Aboriginal community had developed around the Mapoon Mission, established by the Presbyterian Church in 1891. However, in 1954 the church decided to close the mission and, without consulting the community but in collaboration with the state government, embarked upon a process of relocation to New Mapoon that ended in 1964 when police removed the remaining residents. However, from 1974 people gradually returned to the Old Mapoon mission site, settling 'on large spread-out family blocks' that ran along the coastline. In the absence of state oversight and support, they developed a 'strong atmosphere of self-sufficiency and community control', with owner-built houses and subsistence gardens, and re-establishing a close relationship to country, both the sea and the bush.[156] They formed the Marpuna Community Aboriginal Corporation in 1984. In 1992 a Community Town Plan was prepared, but it was rejected by the community because 'it did not recognise the Mapoon lifestyle'.[157] Residents 'expressed a strong desire that Mapoon should not develop into the typical close layout of Napranum and other Aboriginal communities. This requires a shift of focus away from a dense town plan to how Mapoon people live today'.[158] CAT provided that shift of focus.

Moran is low-key about CAT's work with the Mapoon community. 'All we did', he muses, 'was go in and listen to what people were saying and write it down'.[159] Early in 1994 the Marpuna Corporation commissioned CAT's Cairns office, led by Moran, to explore 'how settlement and technology planning can best match community aspirations', and in achieving this goal over the next three years the CAT team recognised that community planning must 'support ... the demonstrated initiative of "owner-builder" family groups. The long term economic sustainability of such "stand alone" groups could be compromised if insufficient consideration is taken to ability to pay for the introduced technology'.[160] The reputation CAT earned at Mapoon resulted in its being given community development work in other communities. At Port Stewart on the eastern coast of Cape York Peninsula, CAT assisted the Lamalama people from 1996 to resettle the traditional lands from which had been forcibly removed in 1961. At Mona Mona, a former mission in the mountains west of Cairns, CAT assisted the Djabugay people from 1997 as they too sought to resettle their traditional lands. In 1998 CAT won an award from the Royal Australian Planning Institute for its inclusive community planning with the Port Stewart Lamalama community.

CAT hoped to apply its experiences in community development in northern Queensland to other remote areas. In November 1999 Walker took part in launching the Desert Knowledge Consortium, a regional development alliance convened at Ross River by Walker as a member of the Northern Territory Research and Development Advisory Committee, Mark Stafford-Smith, head of CSIRO's Centre for Arid Zone Research in Alice Springs, and including CAT, the Alice Springs Council, Tangentyere Council, the Central Land Council, government agencies, the Northern Territory Chamber of Commerce and Industry, and private companies. CAT's board emphasised at the year's end that 'Effective and participatory transfer of appropriate technology can help make progress towards self-determination, sustainable community development, and improved health and well-being for Indigenous people'.[161]

CAT's community development thinking during 1990s evolved beyond acknowledging the need for the mechanics of community participation to address Aboriginal knowledge and culture, and for outcomes to enhance community wellbeing, to recognise that participation and wellbeing in turn depended on the extent to which both could be underpinned by developing sustainable livelihoods for Aboriginal people. Whereas during the 1980s CAT

Participants in the initial workshop of the Desert Knowledge Consortium at Ross River, 1999.

had focused upon developing resilient technologies for Outback Australia and developing consultative processes to ensure that these technologies were appropriate to the communities within which they were applied, in the 1990s CAT increasingly considered the appropriateness of technology in sustaining livelihoods on the homelands. Walker had been alert to this potential during the 1980s, but in 1996 CAT's board made this association explicit by stating that CAT's core goal was the provision of 'appropriate technology services to improve the livelihoods and governance of communities who are remote by virtue of location, access to services, economic resources, political power or cultural difference'.[162]

CAT's experiences during the 1990s demonstrated that this goal would not be achieved merely by developing appropriate technologies and consultative processes:

> CAT's Board and key members of staff recognised that more systematic attention needed to be given to the range of policies, procedures and institutional factors which these problems reflect and which constrain the adoption of lasting solutions to technical problems bearing on a community's health and livelihoods. This was a new and challenging orientation for the organisation.[163]

CAT's collaboration with ATSIC to launch the NTRC was an important new direction. Another was its development of ATWORK to demonstrate how the VET sector could address the education and training needs of remote Aboriginal communities.

CAT's increasing attention to the local employment implications of technology was greatly strengthened by ATWORK. As Walker had declared at a workshop to map out the ATWORK feasibility study in 1987, the homelands movement was creating 'many new opportunities for work and training which hitherto have not existed'.[164] The feasibility study found that almost all major repairs in remote Aboriginal communities were done by non-Aboriginal residents or external agents, and that stereotypes about gender, technology and work had the additional effect of further marginalising Aboriginal women. Community representatives at Galiwin'ku, on Elcho Island, told the ATWORK team:

> Unless government bodies ... change the tendering system and their dependence on outside contractors then local residents will be continually overlooked. In the key areas of housing and construction and repairs and maintenance contractors are continually being used ... The Education Department brings in contractors on $35 an hour to replace taps and washers![165]

At Milingimbi Island, one in the nearby Crocodile Islands group, it was reported:

> There is a desperate need for community-based training for the many capable Aboriginal workers currently employed with council in essential services areas.
>
> They deserve recognition of their skills and a course or courses that will extend their skills to such a point that they can handle routine maintenance in electrical, plumbing and building areas. Until such a course or courses are available in the community, self-management of communities will be beyond the skills bases of the communities.[166]

CAT likewise found during its community planning work at Mapoon that 'Construction by an outside contractor may be "neater" to arrange, but the economic imperatives that drive the construction industry are not particularly amenable to fostering local participation and employment'.[167] CAT's community development work in north Queensland during the 1990s taught the organisation how heavily such remote communities had come to rely on CDEP funding through ATSIC to sustain local development agendas. CAT learned from its early work at Doomadgee 'the importance of CDEP to the economic sustainability of outstations', and throughout the Gulf region at that time, CDEP accounted for almost a third of all employment.[168] CAT's board was well aware of 'ATSIC's recognition of the place of outstations and emerging communities' and of the large flow of funding through ATSIC regional councils for CDEP and housing and infrastructure programmes.[169] ATWORK's training programmes in remote communities were often sustained by arranging with the community councils to fund them through CDEP, the councils identifying the work needing to be done and supplying local CDEP-funded labour with CAT providing trainers to lead and supervise the work. ATSIC's moratorium on funding for new outstations, announced in 1996 while the commission reviewed homelands communities, caused consternation. ATSIC's National Homelands Policy was released in 1999, establishing threshold requirements for funding to new settlements and paving the way for a region by region lifting of the moratorium in the following decade as each ATSIC regional council developed a local plan and had it approved by the commission.

CAT's efforts during the 1990s to develop appropriate technology solutions for Outback Australia, to support Aboriginal empowerment, and to consolidate community development activities that enhanced Aboriginal wellbeing and generated new options for livelihoods, made plain to Bray and Walker one very uncomfortable fact: for all its achievements during the decade, CAT was still a small fish in the big pond of Australian Aboriginal affairs. CAT was now, for the first time, in continuous dialogue with politicians and bureaucrats in Canberra, Darwin, Brisbane and Perth, but the results of that dialogue were uncertain and uneven. A parallel strategy seemed needed: building up the critical mass of research activity, education and training, and public policy advocacy on behalf of the Aboriginal communities of Outback Australia.

****

In 1993 Bray had broached the idea of amalgamating CAT, IAD and Tangentyere Council 'into a college which would create an Aboriginal network' with a central hub in Alice Springs.[170] Bray's suggestion was expanded upon by Walker's 'In Twenty Years Time' paper in 1996. Subsequent discussions broadened to include Batchelor College, near Darwin, which in 1990 opened a second campus at Alice Springs for Aboriginal students. Ongoing discussions among potential partners led late in 1998 to the boards of CAT, IAD and Batchelor College agreeing on principles of association for a joint consortium to establish a DPC, which was intended to be the largest Aboriginal tertiary education and vocational training facility in Australia. IAD urged enthusiastically that the consortium 'express Aboriginal People's aspirations for better education, training, health and greater personal and community empowerment and self determination'.[171] In a show of mutual goodwill, Rose Kunoth-Monks, then Vice Chair of Batchelor Council, joined CAT's board and Kroker became a member of Batchelor Council. Discussions began between the consortium partners and the Northern Territory government early in 1999 to identify a site for the proposed Alice Springs Education Precinct. However, IAD, concerned to maintain its autonomy, withdrew from the consortium in May. Discussions between CAT and Batchelor (which became Batchelor Institute of Indigenous Tertiary Education in July) continued, culminating in a Memorandum of Understanding in September to establish a Desert Peoples Centre. The document was signed by Bray on behalf of CAT, and Gatjil Djerrkura, chairman of Batchelor's council and of ATSIC. It was a promising starting point for the new decade.

Chapter 4

Aboriginal Futures: the 2000s

During the first decade of the 21st century there unfolded an increasingly poisonous national debate amongst non-Aboriginal Australians about the predicted dismal future of remote Aboriginal communities. This debate focused upon supposed moral failure, social collapse and economic paralysis in Outback settlements. Partly in response to these debates and resulting policies, and partly parallel to them, CAT consolidated a set of clear and comprehensive goals for itself as a national Aboriginal science and technology organisation whose core purpose was to achieve a secure and sustainable future for Aboriginal communities throughout remote Australia. As CAT's board noted in 2002, 'technology is the medium not the message ... It is a way of improving a condition'.[1]

CAT's mature strategy platform was made explicit early in the decade with the release of the *CAT Plan 2003–2006: Planning for the Future*. This was based on the vision – so different from those being circulated by the Australian mass media – of 'Happy and Safe Communities of Indigenous People', and it would be achieved by 'Securing sustainable livelihoods through appropriate technology'.[2] CAT's vision and the practical steps for achieving it were realised by its grassroots partnerships with Aboriginal communities and were more broadly communicated in policy documents to politicians and bureaucrats. They were shared directly with Aboriginal and Torres Strait Islander communities via CAT's *Our Place* magazine, and the 'Our Place' radio programme. These engaging fortnightly radio segments, produced by Aboriginal broadcaster Adrian Shaw, were broadcast on over 300 radio stations and radio networks. CAT's alternative interventions in Outback Australia were sorely needed in this generally dismal decade for Aboriginal and Torres Strait Islander Australians.

Planning for the Future was called 'the most important [milestone] in CAT's history'. CAT's chairman, Jim Bray, announced confidently that the plan was 'a fresh statement of our vision and a focus on how appropriate technology

Adrian Shaw interviewing 'Banjo' for 'Our Place' radio.

Our Place *magazine receives wide circulation of ideas and stories on technology.*

can enhance livelihoods'.[3] His linking of appropriate technology to Aboriginal livelihoods and wellbeing drew directly upon CAT's experiences during the 1990s. In Bray's eyes, the dreams that CAT's board had begun to put into words and actions during the 1990s were now becoming concrete realities. Three achievements in particular stood out. CAT's long championship of appropriate technologies for small Outback communities was rewarded at the start of the decade with the launch of its Bushlight solar energy project, funded by a massive injection of federal cash. Likewise CAT's steady advocacy of appropriate water supply systems for Outback communities resulted in important new national partnerships for research on water quality and treatment. CAT's vision of community-focused Aboriginal education and training, to be delivered through a common campus at Alice Springs in partnership with Batchelor Institute of Indigenous Tertiary Education, attracted large-scale funding support from the Northern Territory and Australian governments, and at the end of the decade CAT and Batchelor staff moved into the new campus, the Desert Peoples Centre (DPC).

These achievements were based in part on CAT's now well-established organisational depth and collective confidence, and the significant national influence that came with this. Bray and Walker declared early in the new century that CAT had established 'a national focus as a peak Indigenous technology transfer organisation', by pioneering 'the development of new approaches to sustaining technology and supporting livelihoods in remote Indigenous communities'.[4] New and larger headquarters were opened at Priest Street during 2001, as befitted, in the board's words, the operations of 'a million dollar company'.[5] CAT's accumulated funds then totalled almost $6 million, and it employed 36 people. By 2006, as the Bushlight project gathered momentum, CAT's staff had grown to almost 100 people and the annual operational budget stretched to over $17 million. By the end of the decade, when CAT relocated to the DPC campus, its staff had increased to over 130 and its annual budget to $27 million. The decade also witnessed a seamless transfer of leadership as Bray retired as chairman of the board and was replaced by Peter Renehan, and as Walker stepped down and was replaced as chief executive by Peter Taylor, the former senior manager at the Aboriginal and Torres Strait Islander Commission (ATSIC) who had done so much to encourage CAT's development during the 1990s and to secure federal funding for Bushlight during the early 2000s.

CAT's success also derived from what had become, by the early 2000s, a well-established conjunction between the technical expertise and strategic astuteness of Walker and other non-Aboriginal staff, and the deep cultural knowledge and community ties possessed by CAT's Aboriginal board and general staff. This conjunction generated sustained and constructive interventions in Outback Australia, in contrast to those by the Australian state and also to an extent with those of CAT itself during the 1990s as its board became established, and as Bray and Walker hammered out a constructive working relationship. These new dynamics were clearly evident in the reframing of CAT's mission statement and strategic plan during the early 2000s. The new keyword 'livelihoods', for example, was strongly pushed by CAT's new group manager for its National Technical Services Clearinghouse (NTSC), Steve Fisher, who joined the organisation in 2001 from the Intermediate Technology Group in Britain, and sought to position it explicitly within international community development best practice. However, livelihood was equally a theme that had emerged from CAT's experiences during the 1990s and which expressed Bray's pragmatic commitment to maximising Aboriginal wellbeing. As board members commented, 'Certainly sustainable livelihoods are one end for CAT. Happy and safe communities is another. The Board have indicated previously that terms like self determination are frustrating and do not assist much in a practical sense'.[6] Here was a strongly voiced and firmly anchored Aboriginal perspective upon what, nationally, tended to be a non-Aboriginal debate about Aboriginal policy futures. As Walker said of the board's redrafting of CAT's mission statement, 'the Board had set out the values as a group of Indigenous people who knew their culture and were confident in what they wanted to achieve'.[7]

This Aboriginal perspective, with its emphasis upon 'Happy and Safe Communities of Indigenous People', was informed by direct knowledge of the Outback constituencies within which CAT operated. CAT was well aware of demographic and social trends. Although the number of people living in outstations (communities of under 50 people) was declining, there was overall a rapid increase in the number of Aboriginal Australians living in a broader hierarchy of larger remote settlements. Moreover, although the vast majority of Aboriginal and Torres Strait Islander people in remote areas lived in settlements of 50 people or more, 73 per cent of Aboriginal settlements were outstations and a quarter of Aboriginal people in the Northern Territory lived in such settlements. ATSIC identified 1,216 remote Aboriginal communities in 2001.[8] CAT was aware of the social problems that characterised some of these communities, both small and large. In 2002, for example, when the district medical officer for Docker River made national news for describing Aboriginal children there as being malnourished and anaemic, he was rebuked by CAT board member and Docker River community council president Howard

Smith. However, Smith also expressed his 'shame' at the problems of cannabis addiction, petrol sniffing, poor parental control, poor government support, out of control gambling and an almost total reliance on unemployment benefits within 'what is probably the poorest town in Australia'.[9] CAT contended that poverty and social tensions were symptomatic of the proliferation of remote communities, combined with a 'governance gap' in community decision making and the failure of the Australian state to service them adequately and invest strategically in regional economic development.[10]

However, CAT's experience of working long-term in these 'remote' communities gave it confidence that it could enhance Aboriginal livelihoods there in ways that other service providers could not. Fisher declared that the 'frustration and weariness' evident among the staff of government agencies in their 'constant battle to deliver services in remote areas ... provides an opportunity for CAT to demonstrate that projects can be successful and that an Indigenous organisation can set standards in professionalism and delivery'.[11] Walker contends that until CAT placed attention on livelihoods the concept had been unmentioned in Australian debate about Aboriginal affairs, whereas by the end of the decade it was referred to 'everywhere'.[12]

However, overshadowing local conditions and the delivery of support programmes to communities in need were the accumulating and largely unhelpful effects of a decade of direct federal government interventions in Aboriginal affairs. These were highlighted in 2003 by the Howard government's abolition of ATSIC and its national network of regional councils and, in 2007, by its Northern Territory Intervention. Less newsworthy, but significant in their implications, were the unpicking of ATSIC's National Homelands Policy (the drawing up of which CAT had contributed to) and a federal moratorium on funding housing for homelands from 2006. These policy decisions were not the consequence of a straightforward, conservative political ascendancy in Canberra. Howard's Liberal-National party coalition government had been in office since 1996, but only won control of the Senate in 2005, and in 2007 it was defeated by Labor and Kevin Rudd became Prime Minister. However, Rudd continued the Northern Territory Intervention, and in 2008 devolved responsibility for service delivery in Territory homelands communities to the Northern Territory government, calling into question the ongoing servicing of smaller communities. The Australian government's new policy approach to Aboriginal affairs, released early in 2009 and called 'Closing the Gap on Indigenous Disadvantage',[13] signalled a further round of well-intentioned but paternalistic and assimilationist interventions in Aboriginal communities.

The Intervention and the broader thrust of both Coalition and Labor government policies mirrored the ascendancy in the new century of neoliberal ideas across party political affiliations in reshaping Aboriginal policy. This was expressed in the amalgamation of local community councils into regional shires, winding back the Community Development Employment Program (CDEP), mainstreaming training programmes, reassigning local decision making to private consultants and rationalising government service delivery to remote settlements to a network of larger 'hub' towns. Neoliberalism vulgarised community development thinking, with the result that

> empowerment is conceptualised as participation in local and global markets; institutional capacity building becomes preoccupied with commercialisation; human capital is built through services rather than education; vulnerability is aided by formal legal rights rather than welfare; and citizen responsibilities are cast as moral obligations to community and work.[14]

Neoliberal assumptions provided the essential context within which politicians, academics, and media commentators challenged the ongoing cultural value of the homelands movement, and questioned the moral underpinnings and the socio-economic sustainability of 'remote' Aboriginal and Torres Strait Islander communities. Ralph Folds, a former school principal at Walungurru, mused upon the effects on Australian public opinion of sensationalist television reports and magazine articles, which provided

> the only pictures most people get of the alien settlements scattered in the distant regions of our own 'backyard', and through these upsetting pictures preconceived stereotypes of remote indigenous life are immediately and convincingly confirmed. The settlements look like ghettos and the evidence of degradation seems to be everywhere.[15]

The intellectual underpinnings for this assault were provided in large part by the conservative Bennelong Society (2000–11), a leading figure within which was Peter Howson, who had been Minister for the Environment, Aborigines and the Arts in McMahon's Liberal-Country party coalition government during the early 1970s. Howson questioned not only the viability of Aboriginal homelands settlements, but also the validity of the entire human rights approach to Aboriginal self-determination, which he blamed on the legacy of H.C. Coombs. His arguments influenced Indigenous Affairs Minister Amanda Vanstone in 2005 to brand outstations as 'cultural museums' and to characterise self-determination policies as having 'set up Indigenous people for failure'.[16]

The criticism of Aboriginal communities was not limited to outstations, but covered the whole gamut of Outback settlements. Television stories and newspaper feature articles drew attention to Wadeye, the largest Aboriginal community in the Northern Territory, which was supposedly being turned 'into a war zone' by gang violence.[17] One writer for *Quadrant Magazine* likened Wadeye to a 'failed state'.[18] Such gloomy characterisations were given magisterial confirmation in 2007 by Helen Hughes, a distinguished academic economist who was affiliated with the neoliberal Centre for Independent Studies, in her book *Lands of Shame: Aboriginal and Torres Strait Islander 'Homelands' in Transition*.[19] Late in the decade the hugely experienced and influential anthropologist Peter Sutton delivered a scathing critique of the sense of purposelessness and spiralling violence within many remote Aboriginal settlements, which he argued were symptomatic of a collapse in Aboriginal wellbeing since the 1970s as a consequence of the failure of Coombsian Aboriginal affairs policy.[20]

1. Alternative Interventions

CAT's alternative project was to 'understand what Indigenous people see as viable for their livelihood not what external agents say should be viable'.[21] CAT's board members rejected mainstream media sensationalism about Outback settlements and opposed the thrust of federal and state interventions in Aboriginal affairs. They also expressed dissatisfaction with the assumptions built into key non-Aboriginal concepts that shaped the new direction of Aboriginal policy. They especially distrusted the preoccupation with 'remote communities' and with basic terms that had conditioned Aboriginal policy during the 1970s, 1980s and 1990s. They considered that 'the concept of self-determination is defeatist in that [our] people will always be a minority group'. They characterised empowerment as 'a one way street. To suggest one group of people need empowering is to admit that these people have no power to begin with'.[22] They questioned the 'practical limitations of a rights and equity based agenda', arguing that this approach doomed its non-Aboriginal proponents to 'work within the welfare dependency, supply driven developmental model that contributes to the current malaise'.[23] They complained that they saw 'a lot of humbug from people who believe they should get a service purely because they are Indigenous', and proposed that the board 'move away from a welfare based approach to services towards a service that facilitates access rather than supply services. It requires people to demonstrate or acquire capacity and to adopt good governance practices'.[24]

The board sought to make this fundamental policy change by turning whitefella thinking on its head: 'Instead of going into a community and saying "what's wrong with this community" we should be saying "what things are going [well] with this community'".[25] Instead of dwelling upon rights and dependence, they emphasised the 'wellbeing' theme that CAT had identified during the 1990s and sought to expand upon it by delving further into the concept of 'sustainable livelihoods', which the board defined in 2001 as the 'range of activities that support improved well-being through work, enterprise and trading and that can be maintained into the future'.[26] Consolidating livelihood opportunities entailed engaging with remote communities rather than despairing of them. It meant 'emphasis[ing] the strengths of people, rather than their needs'.[27] It highlighted CAT's core purpose, to 'bring about change ... Are we the voice for Aboriginal livelihoods? Are we a voice for the future?'[28]

The fundamentals of CAT's alternative approach to Aboriginal affairs were hammered out at two workshops in 2001 and 2004 that the board and senior management held at Glen Helen, 140 kilometres west of Alice Springs in the West MacDonnell Ranges. The first meeting at Glen Helen, held over 27–29 July 2001, represented the culmination of discussions since 2000 about refining CAT's vision and purpose. It was at Glen Helen that CAT's vision statement was rewritten to focus upon 'happy and safe communities of Indigenous people', and its goal of achieving sustainable livelihoods first defined and, with Fisher's input, given a firm methodological basis.[29] It was also here that the strategic decision was taken to direct attention to the increasing numbers of Aboriginal people living in communities of 200 people or less, and to harness 'appropriate technologies in securing sustainable livelihoods' by providing technical support and technical services, and – a new direction in CAT thinking that built upon experiences during the 1990s – by developing opportunities for enterprise and trading.[30] The decisions taken at Glen Helen were reported back to CAT staff, further debated and amended, and the final wording of the vision statement settled by the board in August 2002. The parallel process of drafting the 2003–06 strategy document, *Planning for the Future*, was completed at a follow-up workshop in April 2003.

This refocusing process resulted in a simplified and more effective organisational structure, with activities divided into three main groups: research and outreach activities were concentrated in the NTSC; and other core activities were consolidated into Education and Training, and Corporate and Policy groups. More important than organisational structure were the activities delivered. *Planning for the Future* emphasised that CAT now offered 'a range of [appropriate technology] services in education, training, research, technologies and products, media and policy work', all designed to increase the knowledge and skills of Aboriginal people and their active participation in networks and partnerships, and moreover tailored to 'meet [the] particular economic, cultural, environmental and social needs of people'.[31] This core approach, and the associated goal of achieving sustainable livelihoods, were comprehensively explained in the NTSC's revamped *Our Place* magazine, published three times a year; a new 'Our Place' series of national radio broadcasts, started by Adrian Shaw in 2002, which delivered 'positive stories from remote Indigenous communities about sustainability';[32] and BUSH TECH technical notes (published in *Our Place* and subsequently made available for internet download) for general readers on diverse topics, for example, water harvesting, hot water, renewable energy systems, telephones, radio networks, rubbish tips and maintenance advice about roads, septic tanks and water bores.

The second Glen Helen workshop was held over 14–15 July 2004, and resulted in a further overhaul of CAT's thinking and operations. The workshop was triggered by two developments. The first was a major external review of CAT, conducted by Rick Callaghan, an Aboriginal consultant and the director of Adelaide-based Yaran Business Services, undertaken by its major funding partner ATSIC in 2002–03 as a follow-up to its earlier review in 1995–97. Callaghan's report and recommendations were released late in 2003. The second development was the abolition of ATSIC by the Australian government in 2003. An urgent rethink of CAT's fundamentals was clearly needed, and the theme for the second Glen Helen workshop therefore became 'CAT in Transition – A Structured Refocusing of the Organisation'.[33] The most important consequence of this reassessment was a revision of CAT's strategic thinking in order to address Callaghan's recommendation that CAT

> increasingly focus on impact and outcomes rather than outputs. This relies on a closer management of relationships and our relevance to communities and their changing circumstances and a need for CAT to capture the key learnings from its practice and to be at the cutting edge of thinking in Indigenous development.[34]

* * * * * * * * * * * * * * *

Callaghan's review pinpointed an underlying feeling within CAT's community that its development activities urgently needed reenergising. Board members conceded in 2002 that there 'was a perception that communities had moved on but CAT hasn't'.[35] A case in point was the old workshop where CAT's homelands activities had begun during the early 1980s. Fisher tabled a review of the workshop early in 2002, acknowledging that it had played a central role in assisting remote communities since its establishment in 1980, but concluding that in recent years it had lost its way. Sales of workshop products had declined sharply after 1996, before stabilising at a lower level from the early 2000s (mostly through sales of drum ovens, stoves and chipheaters, together with crowbars, bush tucker digging sticks and some VIP toilets). Fisher argued that as 'the pace of change in CAT has quickened … there has been a marked shift in the nature of work undertaken by the organisation towards research, technical services and the provision of information'. With innovation occurring in these other areas, 'a certain haziness within the organisation around the reasons for the existence of the workshop' had developed, and Fisher suggested that this mood was reflected in conditions within it: 'Any visitor to the CAT workshop is struck by the apparent clutter and disorganization … The edges of the workshop space are a dumping ground for old machines and unfinished jobs, some of them long out-of-date'.[36]

By contrast, some of CAT's most important work during the 2000s took the form of service delivery projects in remote areas, which resulted in the opening of additional coordinating regional offices in Darwin in 2003 and in Kununurra (Western Australia) during 2005. CAT's activities in the north-west of Western Australia, which had begun on a small scale during the late 1990s, snowballed during the early 2000s under the direction of CAT's regional manager Marc Seidel. Funding from ATSIC and the state government enabled CAT to provide technical advice, construct and maintain roads and river crossings, upgrade and maintain airfields and firebreaks, and manage civil works for outstation communities throughout the Malarabah and Kullarri regions in the Kimberleys. These operations would expand right across Western Australia's 'Top End', to Kunamurra and Wyndham. Similar community partnerships were developed later in the decade with the network of Aboriginal communities along the Plenty Highway, between Alice Springs and north-western Queensland. Other important community development work performed by CAT was settlement planning in north Queensland, building upon its pioneering work there during the 1990s, which CAT proudly described as being a model for community planning. At the Dawnvale pastoral lease on Cape York, for example, following its return to its traditional owners, CAT worked with the Bana Mindilji people to develop and implement a livelihoods and local management plan. CAT staff also assisted the Kuku Yalanji people as they returned to the old Buru pastoral lease in the mountains inland from Cape Tribulation.

In a new application of CAT's overlapping expertise in community development and appropriate technology research, the organisation became an associate partner in the federally funded Cooperative Research Centre (CRC) for Water Quality and Treatment in 2000, and in 2002 became a full research partner, winning new externally funded staff positions as a result and undertaking water research projects in Queensland, Western Australia, the Northern Territory and South Australia. When the CRC's start-up funding ran out it re-established itself commercially as Water Quality Research Australia in 2008, with CAT a founding research partner. CAT's participation in CRCs broadened when in 2002 the organisation became a partner in the Desert Knowledge CRC that established its headquarters in Alice Springs. It was led by CSIRO scientist Mark Stafford -Smith, with whom Walker had collaborated in launching the Desert Knowledge Consortium in 1999.

CAT's successful partnership with Rio Tinto, begun during the 1990s, was extended in 2006 for a further three years. The partnership sustained two main activities. The Rio Tinto annual fellowship programme applied the skills of Rio Tinto staff to assist CAT in a wide diversity of tasks in 'remote' communities, ranging from water supply and transport technology to housing, energy use, telecommunications, environmental health and communicating through drawing. Seidel recalls that in the Kimberleys he had Rio Tinto engineers on CAT fellowships 'busting their guts' to design and install river-crossing culverts, composting toilets and other infrastructure.[37] The annual Akaltye Youth Event (Akaltye is based on the Arrente word meaning 'opportunity to learn') brought together Aboriginal junior secondary school children for a week of science and

The Ngallagunda Crossing in the Kimberly of WA designed by CAT Rio Tinto fellows.

technology education. First held in Alice Springs, in 2005 it took place at Weipa on the Cape York Peninsula and in 2007 at Gladstone on Queensland's central coast. Dale Jones, an engineer from Western Australia who joined CAT's board in 2008, was introduced to science by the Akaltye Youth Event, and later undertook a Rio Tinto fellowship with CAT.

The applied research that such partnerships and funding streams maintained, continued to produce appropriate technology products that supported Aboriginal wellbeing and livelihoods in remote Australia, regardless of the overall thrust of neoliberal policy. As Robyn Grey-Gardner, a water researcher at CAT, remarked of CAT's collaborative work on water quality, 'you can't package it into something that we can just deliver'. This was work that generated a social benefit, not a profit.[38] Collaborative research in the field of communications resulted in CAT designing a robust public phone that would survive severe climatic conditions and heavy use. Conventional public phones did not adapt well to the Outback: coin-slots would jam because the coin boxes were rarely emptied, keypads would clog with sand, the telephone cord would break and the handset often fractured. The new community access phones were successfully trialled in 20 locations during 2005, and late in the year an agreement was signed with Telstra to take over CAT's patent (which won a design innovation award during the year) and install the phone in 200 remote locations, using CDEP funding to train residents in its operation and maintenance. Taylor conceded ruefully that 'CAT learned a big IP lesson' in signing over the patent to Telstra, but CAT was later contracted by the Australian government to service the phones because no one else would tender for the job.[39]

****

The clearest example of CAT's efforts to support Aboriginal livelihood choices in small Outback communities is CAT's Bushlight solar energy project that began in the early 2000s in partnership with the Perth-based Australian Centre for Renewable Energy (ACRE). Bushlight combined applied research, product development and service delivery to meet the energy needs of the many remote Aboriginal communities that were not connected to the electricity grid and depended on either costly generators or unreliable solar systems. A study in 1997–99 of renewable energy systems in 88 Aboriginal communities found that approximately one third of the systems were not operational, that insufficient system capacity meant that many other systems were unable to cope with peak demand and that residents were 'disgruntled' with this state of affairs.[40]

CAT had become a partner in ACRE when the cooperative research centre was established with Australian government funding in 1997. Led by Walker and Taylor, CAT and ATSIC organised a major Energy Forum in Alice Springs during 1999, the success of which led ATSIC to propose to the federal government an ambitious scheme to use diesel excise under the new GST to deliver reliable solar energy to small and remote Aboriginal communities. Taylor lobbied for the proposal in Canberra, targeting not only the government, but also the Australian Democrats who held the balance of power in the Senate, and early in 2001 it was confirmed that CAT and ACRE would receive $2 million per annum over four years (from July 2002 to June 2006) from the Australian Greenhouse Office (AGO) to ensure adequate community consultation and planning occurred before RE systems were introduced. The AGO funding was matched by $2 million per annum over four years from ATSIC in a dollar for dollar arrangement to supply capital for RE systems as part of the Bushlight programme. The Howard government supported the initiative because it wanted to send a 'clear message ... that [it should be] possible for Indigenous people to live wherever they can [and still] enjoy the benefits of Australian citizenship'.[41] Bushlight was expected to deliver reliable energy services to 200–240 communities across the Northern Territory, Western Australia, Queensland and South Australia, most of them of less than 50 people and not connected to the electricity grid.

An unincorporated joint venture called Bushlight was established with ACRE early in 2002 to deliver the programme, and additional staff were appointed in Perth, Alice Spring, Cairns and Derby. Bray chaired the board of the new company, Walker became its executive director and Grant Behrendorff was appointed general manager. Behrendorff had previously worked as a power company manager in Queensland, where he had been responsible for delivering power to Aboriginal communities on Cape York. The idea was raised of developing the joint venture into a commercial undertaking, profits from which could sustain Bushlight once the four-year start up funding ended. A complication arose late in the year when ACRE's application for renewed funding was unsuccessful, leaving CAT the sole partner in Bushlight and prompting CAT's board in early 2003 to reassess the viability of the new venture. In August 2003, as ACRE was wound up, the board agreed to make Bushlight a fully operational unit within CAT. The board also began negotiations to form a commercial joint venture to be called Desert Services Australia. However, this early interest in developing a commercial approach did not take hold formally until some years later.

During the first year of operations, from mid 2002, the first generation of Bushlight solar systems was designed and over 100 communities were visited to discuss their energy needs. By December 2004 Bushlight teams had visited over 430 communities across the Northern Territory, Western Australia and Queensland, and were working with 200 communities to design and install Bushlight systems. (Local stakeholders in South Australia advised that renewable energy systems there were already in place, and Behrendorff decided that the Bushlight team was already 'so busy that we [didn't have] to go into places where they didn't want us').[42] Community members received basic training in system use, and Bushlight developed a nationally accredited Certificate III training course in renewable energy system maintenance. CAT's 2004 workshop at Glen Helen recognised Bushlight as CAT's best opportunity to establish itself nationally as a 'viable integrated technical service provider'.[43]

With the first phase of federal funding due to end in 2006, CAT increasingly turned its attention to transforming Bushlight into a commercially sustainable operation, and to gauging whether by nesting Bushlight services within a Desert Services Australia model, now to be known as Integrated Technical Services, this department could become CAT's commercial arm. During 2007, with a funding boost of $7.7 million in federal funds provided by the Department of Families, Community Services and Indigenous Affairs (FaCSIA) over another three years, CAT employed over 100 people on Bushlight operations and extended the programme to another 120 communities. Bushlight won a National Engineering Excellence Award from Engineers Australia in 2007, and Behrendorff won the Institution of Engineers' technologist of the year award in 2008. During the previous year, with further funding support from the Australian Greenhouse Office, Bushlight staff also began a trial project in India, working in villages in West Bengal and Orissa. Looking back on CAT's achievements in renewable energy for remote communities during the decade, Behrendorff declared proudly in 2010 that Bushlight had 'really hit it'.[44]

2. Risk

The effectiveness of CAT's alternative interventions was conditioned, however, by three complicating factors. The first of these was the challenge of maintaining cohesion and focus during a time of rapid growth and fundamental organisational change. The second was the headache of how to ensure organisational good governance and leadership renewal. The third was the continuing dilemma of sustaining programmes and community trust while remaining dependent on third-party funding by government.

The pace of CAT's growth during the decade exposed lagging inefficiencies and generated internal tensions. CAT's stocktake at Glen Helen in 2004 was prompted in part by concerns that CAT's increasing size had not been matched by improvements in the organisation's culture and organisation. The workshop resulted in an updated constitution (approved in 2005), a streamlined board of Aboriginal directors, reconceptualisation of the Director's role into that of a Chief Executive Officer (CEO), and the establishment of a new Chief Operations Officer (COO) position to manage general operations. However, beneath the surface, with the continuing increase in staff numbers and the scale and diversity of operations came uncertainties and frustrations over purpose and structure, as well as the direction of change. Notwithstanding the apparent confidence with which the board had stated CAT's vision and goals early in the decade, Walker was privately dismayed that during the associated consultations with staff 'it was not clear that all staff had a consistent view of what was involved in appropriate technology or the breadth of issues that could be encompassed'.[45] By 2007 board members also expressed concern that 'we have lost sight about what CAT was in the beginning and we need to get it back, it's like our own identity that no-one else has'.[46]

The uncertainties that had surfaced during the 1990s about CAT's character as an Aboriginal organisation also continued into the new decade. In 2000 board member, William Lane, advocated 'a framework to allow Indigenous staff to progress to senior levels through training and mentoring'.[47] The issue of CAT's 'Aboriginality', and the possibility of recruiting an Aboriginal assistant director 'to groom as [the] Director's successor', were discussed at a board workshop in 2002.[48] A follow-up board workshop in 2003 resolved to search for an Aboriginal deputy director, and if this were unsuccessful, to offer Fisher a new position as managing director 'to mentor [the] development of an Indigenous Deputy Director'.[49] However, the same workshop also confirmed that:

> CAT does not have an Aboriginalisation policy. The Indigenous Board is comfortable that the organisation is controlled and run by an Indigenous Board in terms of Aboriginalisation. Given that CAT is unique in research, training and delivery in disciplines of science and technology it is often difficult to find Aboriginal people with *both experience and qualifications* in these disciplines, it is equally important to acknowledge that the organisation selects staff based on their merit with strategies in place to train/mentor Aboriginal candidates if not selected on their merits outright. As the business of CAT requires an experience of Indigenous and non Indigenous life it is important for the work of CAT that the prevailing values represent the concerns and interests of Indigenous people but at the same time provide for a sharing of knowledge across cultures.[50]

Bray felt comfortable that CAT's Aboriginal board could 'ride on the … backs' of its non-Aboriginal staff.[51] Nonetheless, the board cautioned in 2007 that notwithstanding CAT's increasing number of staff, it was employing relatively fewer Aboriginal staff.[52]

Fisher's tenure as group manager of the NTSC and then deputy director between 2001 and 2005 brought internal tensions about vision and direction to a head, but served to channel them into largely constructive discussions and positive change. Fisher acted aggressively as an agent for renewal within the organisation. Late in 2001, just three months into his appointment, Fisher released a review paper and set of recommendations for the NTSC which quickly became known as his '100 Days at CAT' manifesto. Fisher's paper, which amounted to a de facto review of CAT's core operations, triggered a series of meetings and sometimes heated discussions among the board and staff that continued through most of the following year. Fisher warned that CAT's once distinctive identity had become diluted, and that they must work hard 'to overcome the perception that CAT is simply another public sector organisation'. Fisher also suggested that CAT's overall sense of strategic direction had become 'muddled', and that its programmes 'therefore give the appearance of being a loose agglomeration of activities for which money could be obtained rather than work which is driven by an organisation making sound strategic choices'. Fisher suggested that there was insufficient monitoring and evaluation of CAT activities and that underperformance and absenteeism amongst staff was much higher in CAT than in equivalent organisations.[53]

Fisher ruffled more feathers in 2002 when he recommended that CAT's production workshop be closed, arguing that its neglect over time 'implied that really the manufacture of goods is something from the past and that CAT's management focus has moved firmly towards research, integrated services and knowledge management'.[54] Walker agreed, complaining that despite CAT's expanding activities 'Many people [still] see CAT as being just chip heaters … and VIP latrines and technical training'.[55] Some on the board clung to those past achievements, pointing out that 'a lot of people still associate the products with CAT'. However, board member Victor Hunter responded that the production workshop encapsulated 'the welfare mentality of CAT' that needed to be overturned in order for the organisation to move forward.[56]

Alongside concerns about CAT's structure and purpose, anxieties over its governance practices simmered, and increasingly so over succession planning for its senior managers. Attention focused upon the board's performance and upon sharing (and ultimately transferring) the burden that fell on Walker.

Even at the beginning of the decade, board members fretted about the amount of business that was being squeezed into their meetings and, as the scale and complexity of CAT's activities increased during the decade, about board members' skills and time commitment to keep abreast of them. Walker questioned 'the degree to which the Board of CAT understand and control the now complex operation that has grown over the past few years and whether the mix of Board members has the necessary breadth of skills to maintain control of the organization'.[57] The board sought to recruit both community representatives and high-profile Aboriginal leaders and to ensure their adequate induction and training. At the Glen Helen workshop in 2004 much time was spent discussing how to improve the board's performance and recruit young directors who were endorsed by their respective communities and who also shared 'a passion for [CAT's] vision'.[58]

CAT also wrestled with the issue of how to manage Walker's workload as the organisation grew in size and influence. The 2001 Glen Helen workshop pondered how to free the Director from the time-consuming distractions of day-to-day management and considered dividing management responsibilities between a COO in charge of day-to-day operations and a CEO with responsibility for strategic planning. Fisher effectively began to act in the former role after joining CAT in 2001, and especially after the board decided in principle in 2003 to make Walker CAT's executive director and Fisher his deputy. Putting this principle into practice was, however, complicated by the related issue of succession planning. As Fisher had commented in his '100 Days at CAT' paper in 2001, CAT was 'over-dependent' on Walker because 'much of the [organisation's] knowledge … is stored in his head'. Fisher warned that the 'result is a real vulnerability since to many agencies in Australia Bruce Walker *is* CAT'.[59]

CAT's board brought in consultant Gregor Ramsay in late 2002 to run a workshop on succession planning. The looming problem was clear: Walker had led the organisation since its inception and was now aged 53; a renewal of his contract would take him to retirement age. Likewise, Bray was already in his sixties and had been CAT's Chairman for over 12 years. The challenge in replacing both men was also obvious: how could the board be confident that in replacing them CAT 'remains CAT'.[60] The Glen Helen workshop in 2004 further agonised over these issues and restated the 'need for a transition of leadership and a consolidation of new opportunities within the organisation'.[61] The workshop resolved to abolish the old position of Director and the imprecise new position of Deputy Director and instead to reappoint Walker as CEO and Fisher as COO or General Manager. When Fisher left CAT in the following year to work at the Desert Knowledge CRC, Walker resumed both roles but warned the board that although 'I am enjoying getting back into operations [it is] not sustainable', and he pointed to 'signs of some fraying around the edges' of strategic planning and day-to-day management.[62]

Peter Taylor, who had joined CAT as research manager after Fisher left in 2005, was appointed CAT's COO in 2006 and became acting CEO in the following year when Walker took five months' study leave overseas. It was decided in 2008 that Taylor would formally take over as CEO when CAT moved to the Desert Knowledge Precinct. The succession occurred in January 2010. Bray retired as Chairman in September 2010 but remained on the board.

****

Despite CAT's development into a large-scale organisation during the early 2000s, it remained highly vulnerable to decision making within government about key funding allocations. The early development work for Bushlight, for example, was carried on against a backdrop of anxiety about the outcome of ACRE's application for renewed CRC funding, and when this was unsuccessful further anxieties were generated by the need to make Bushlight commercially sustainable once its initial government funding ran out. At Glen Helen in 2004 attention turned to generating marketable intellectual property through Bushlight and to selling solar energy services through Desert Services Australia. CAT's board had resolved in 2000 that the organisation should aim to generate a quarter of its revenue from commercial activities. In 2007 CAT's constitution was amended to allow it to establish a wholly owned commercial company, and in the following year Ekistica Pty Ltd, trading as CAT Projects, was launched to undertake commercial projects. The new company simultaneously launched a trial Bushlight project in India, funded by the Australian government, and undertook the design and management of two solar energy projects in Alice Springs as part of its participation in and leadership of the Alice Springs Solar City project. One project was a solar array at the Crowne Plaza complex, the largest roof-mounted solar system in the southern hemisphere. The other was the Solar Centre (opened by the federal Minister for the Environment, Peter Garrett, in 2008) at the Desert Knowledge Precinct where Desert Knowledge Australia and the Desert Knowledge CRC were based and the DPC was being built.

Commercialisation was designed to reduce the vulnerability of CAT's activities to decisions made by government funding agencies. The viability of CAT's Cairns office, for example, remained questionable during the early 2000s because of government hesitation to support its operations. In Western Australia, Seidel's community development work was jeopardised when in

2006 FaCSIA withdrew its funding for CAT's civil works programme in the Kimberleys. Seidel fumed that 'after six years of very useful and very successful programming in improving access and safety for all those little communities out there with a regular cycle of airstrip maintenance and road construction and road maintenance, it was pulled overnight'.[63] In 2007 FaCSIA went still further and terminated its funding (begun early in the decade) for CAT to deliver basic municipal services such as rubbish collection, airstrip inspection, firebreaks and local maintenance works in the Kimberleys. The programme, initially providing for 10 Aboriginal communities, had expanded to almost 50 communities throughout the Kimberleys when FaCSIA withdrew its support. Walker, looking strategically at all the programmes that CAT had built up in remote Australia, worried that the Australian government's devolution of responsibilities in Aboriginal affairs to the state and territory governments during the decade would make CAT's core activities 'really difficult'.[64]

Behrendorff speculates that Bushlight's success was attributable in part to a unique set of circumstances that resulted in CAT being able to set up the programme without government oversight and interference:

> we didn't have to go back singing for our dinner every six months and try to justify our existence because a lot of projects you get involved with you spend as much time stakeholder managing as you do actually trying to achieve the job ... ATSIC became ATSIS and then ATSIS became DIMIA and then DIMIA became FaCSIA ... Basically the nett effect of that was that nobody really had much involvement in the project from a funding perspective ... and very few questions [were] asked. The secret I think to successful outcomes for Bushlight was that we had four years before anybody took a critical look at us.[65]

The normal experience for CAT staff was having to 'flit' from one funding agency to another, according to what was currently available, and having to deal with an 'inevitable change of personnel half way through' the funded project.[66] CAT senior manager Metta Young estimated that 'half of my work is networking'.[67]

Dependence on government funding jeopardised not only CAT projects, but also its credibility with community partners. Walker asked the board in 2002: 'How can we get input and provide feedback to remote communities to ensure we have a balance between agendas set by funders and external stakeholders and communities?'[68] Board members experienced this problem at first hand when they visited Western Australia in 2003. At Derby and Fitzroy Crossing they faced accusations by community representatives that CAT 'had muscled in' on government grant money and that 'CAT had been at fault'. As the delegation noted, the 'discussion at times got as hot as the weather'.[69] Walker restated the broader problem at CAT's annual general meeting in 2004, where he warned that there would always be difficulties in 'services delivery because of programs/systems that are imposed upon CAT from outside sources. There will always be a clash for community versus the outside'.[70] In 2007 Taylor described as 'disappointing' the advice received from FaCSIA bureaucrats that the Australian and Western Australian governments intended to discontinue CAT's municipal service delivery operations in the Kimberleys. Taylor told the board that it had become 'extremely difficult to maintain a good working relationship with FaCSIA [because] their decision making has increasingly centralised and become closer to the Minister'.[71]

Changes in government policy, in addition to endangering CAT's financial planning and community partnerships, challenged its fundamental principles and goals. Fisher cautioned in his '100 Days at CAT' paper that the 'issue of policy is perhaps the most challenging for us to handle', because CAT sought to maintain a position of political neutrality in what was becoming an intensely politicised policy environment. As Fisher put it to his colleagues, the 'issue for CAT is whether or not we advocate policy change and in what way'.[72] CAT could watch quietly from the sidelines as the outcomes it had worked to achieve for over 20 years were undermined, or it could explicitly advocate on behalf of Aboriginal communities and risk the consequences.

The dilemma facing CAT became clear in 2003 with the abolition of ATSIC, CAT's major source of funding and its main mechanism for recruiting Aboriginal representatives to the CAT board. Paul Wand, a vice president at Rio Tinto and the architect of its partnership with CAT, noted sympathetically that CAT's management became 'a bit desperate' upon ATSIC's demise.[73] The biggest fallout for CAT was the ending in 2004 of the ATSIC funding stream that had sustained the NTSC. The national clearinghouse was replaced by a scaled-down but still energetic research division, the Technical Resource Group, which was headed by ex-ATSIC manager Taylor.

However, it was the emerging neoliberal consensus over laying blame upon remote Aboriginal communities that most perturbed CAT. The Glen Helen review of CAT's strategic plan in 2004 was largely performed in response to ATSIC's abolition and to the 'global changes in terms of Indigenous affairs' that CAT recognised were under way.[74] Walker referred in 2006 to the 'uncertainty[,] despair ... and confusion over the public utterances of the Minister [for Indigenous Affairs, Amanda] Vanstone and others around the viability of outstations'.[75] In the following year CAT was further stunned by the Australian government's Northern Territory Intervention, and late in the year by the decision of the incoming Rudd Labor government, when meeting with the state and territory governments at the Council of Australian Governments (COAG), to establish a new unified approach for addressing Aboriginal disadvantage by concentrating service delivery for remote communities in regional hubs. At CAT's annual general meeting in 2007 Walker acknowledged 'the rapidly changing environment and the confusion created by the government changes', and Taylor spoke grimly about the 'difficult time' that the organisation now faced.[76]

CAT's response to changing government policy and public perceptions was to reassert boldly the homelands focus with which CAT had begun, and to anchor this within the broader hierarchy of settlement types in Outback Australia rather than servicing outstations alone. The Glen Helen workshop in 2001 reaffirmed the board's decision, made at a workshop in 1999, that 'CAT's core business is appropriate technology transfer for remote Aboriginal communities, and niche opportunities to meet priority client needs',[77] and later in the year Fisher's '100 Days at CAT' defined the 'core ... work of the NTSC' as being to provide the 'opportunity for people to move back to their lands'.[78] Bray restated this position at CAT's annual general meeting in 2003, drawing attention to the livelihood needs of small communities of under 200 people.[79] The board responded to Vanstone's comments about Aboriginal homelands by reiterating CAT's vision of 'Happy and Safe Communities' and emphasising that this 'Vision seems to be more urgent given the current government decisions around remote communities/homelands'.[80] The board's position strengthened still further after the Intervention. Notes taken at the September 2007 meeting of CAT's board record the members' mood: 'CAT to continue with the core business and look after people on homelands'; 'Tell a positive story that it is possible to live out on homelands in a healthy and safe lifestyle'.[81]

In 2005 CAT explicitly addressed the changing mood of government policy and community opinion regarding small remote communities by revitalising its early Bushlight umbrella organisation, Desert Services Australia, renamed Integrated Technical Services (ITS), its task being to

> pilot and progressively monitor and evaluate linked processes of community engagement, planning and delivery of essential services (shelter, water, waste and energy) on an integrated demand responsive service model for targeted remote Indigenous communities, so that they are provided sustainable (i.e. affordable, consistent and reliable) services. ITS will also improve community capacity to choose and manage their technical services by

> providing communities with information, training and support services, so that communities and their service providers are better equipped to manage and maintain their services.[82]

CAT acknowledged that

> Over recent years, national policy interest has increasingly focused on the sustainability of small remote Indigenous communities, especially outstations. Government has questioned whether the cultural attachment to country facilitated by outstation settlement comes at the price of highly inefficient or non-provision of services, and that outstation residents are placed at risk of ongoing health, education and economic disadvantage.

ITS was intended to turn this thinking around by building upon the Bushlight experience and developing 'better mechanisms ... to provide reliable basic services to remote Indigenous communities'.[83]

Paradoxically though, the initial two-year resourcing of ITS's outstation servicing project was made possible by Australian government funding. This was tied to regular review by Commonwealth officials and the achievement of specified outcomes. CAT's alternative interventions in remote Australia therefore carried the immediate risk of making the organisation even more dependent upon government funding. Moreover, CAT's emphasis on small communities in remote Australia carried the broader risk of its being marginalised, along with its community partners, from the main thrust of government Aboriginal policy and from the general public's good opinion. The risk of marginalisation became clear in 2008 with the announcement of two new policies in the Northern Territory. Firstly, the Northern Territory government amalgamated community councils into larger shires, increasing Aborigines' frustration at the erosion of their ability to participate in local decision making and to hold officials to account. 'We got no power since the Shire come in', said Hayes, and he worried that the new managers 'don't get involved with the people or anything'.[84] Secondly, the Northern Territory and federal governments formally agreed to transfer responsibility for homelands policy to the Territory government from mid year, and also decided that no federal funding would be provided for housing on outstations. Both decisions carried the risk of distancing CAT as well as its community partners from decision making and funding allocations.

ITS was a casualty of these developments. Dependent as it was on outsiders' agendas, its operations became remote from the communities it was intended to serve. Walker admitted that it 'fell down a bit', and that it highlighted the risk of 'moving away from your roots'.[85] Moran more bluntly contended that it had failed because it was a product of 'people in a room with a whiteboard' rather than having its 'genesis [in] someone going into a community'.[86]

The broader risk to CAT as a result of changing government policy became still more evident in 2009. In January COAG's new position was formalised through a National Partnership Agreement on Remote Service Delivery, which undertook to concentrate 'whole of government' activities into service hubs located in the 'larger and more economically sustainable communities' which could be accessed by smaller surrounding settlements.[87] In February 2009 the Rudd government spelled out its new approach in the policy paper *Closing the Gap on Indigenous Disadvantage*. An important element in the Australian government's new policy was the appointment of a Coordinator General for Remote Indigenous Services, who would be responsible for efficient service delivery in 26 'priority communities' in Remote Australia.[88] CAT's board watched with misgiving the Northern Territory government's release of its complementary *A Working Future* policy statement in May 2009, which included an undertaking that 20 of the Territory's 'biggest remote communities will become proper towns, with services, buildings and facilities like any other country town in Australia'. The new policy's subtext was that smaller homelands settlements would need to access support services through these 'proper towns', and that no government funding would be available to establish new homeland settlements.[89]

****

The implications of this kind of policy approach for CAT's operations could already be seen in the organisation's reluctant winding back of its education and training activities. CAT admitted in 2003 that 'Education authorities are implementing a national "one size fits all" training system driven by mainstream industry. Demand driven and effective outcomes for communities are increasingly difficult to implement'.[90] ATWORK, which had been launched with so much enthusiasm in the early 1990s, became a victim in the early 21st century of government emphasis upon mainstream trades-based training in the TAFE sector and of shifting funding priorities in vocational education away from CAT's core objectives. Metta Young recalled that these changes were 'a killer' for ATWORK.[91] CAT's Education and Training Group (renamed the Technical Skills Group in 2005) struggled to meet its targets and in 2002 the board decided that 'existing remote programs be wound up or wound back'.[92] Notwithstanding this consolidation, enrolments continued to decline in ATWORK's Certificates I, II and III in Applied Design and Technology, as well as in CAT's other certificate programmes in automotive servicing and general construction skills; more and more of the remaining trainees dropped out of their courses. This problematic retention rate did not necessarily mean poor training outcomes at a local level, but it left CAT out of step with government regulators. As CAT noted in 2002,

> CAT students, particularly students in remote communities, complete their community work project but do not necessarily complete a qualification. The concept of 'industry' cannot be easily transported to discrete Indigenous communities. The national focus on mainstream industry groupings and employment outcomes has limited relevance to developing technical capabilities and improving livelihoods of communities of Indigenous people.[93]

In 2006 Australian government regulators required CAT to overhaul its programmes and upgrade the formal qualifications of its teaching staff in order to be reaccredited as a registered training organisation. During the year Robyn Ellis, who had done so much during the 1990s to build up ATWORK and the associated Women and Technology Network, resigned dispirited from CAT. She felt that the 'living issues' in remote communities that ATWORK had been designed to address – local design and maintenance issues, for example, and rubbish management and erosion control – could no longer be addressed by the new and emasculated ATWORK programme. Vocational Education and Training (VET) requirements had, in her opinion, become 'a treadmill' for educators at CAT.[94] Another departure from CAT in 2006 was Ron Talbot, who with Kurt Seemann had first designed the ATWORK curriculum. Talbot recognised that only a 'smattering' of the original ATWORK concept remained in the curriculum, and that in consequence he had 'lost some of the fire in the belly'.[95] Walker expressed frustration at the 'sense of paralysis' clouding CAT's education, training and employment initiatives as a result of these shifting public policy emphases.[96] Seemann pointed out that the Australian VET sector still 'refuses to accredit locally brilliant Aboriginal technical knowledge' as a legitimate element within 'a wider more conventional suite of skills'.[97]

CAT's decision to push ahead with the DPC can be seen in part as a reaction to this crisis in CAT's education and training programme. It was a reaction too, to the broader risks that faced CAT in an unsympathetic policy environment. The DPC was designed to be 'a catalyst for change in the desert'.[98] It sought to turn things about fundamentally in the grassroots of Outback Australia. The DPC concept was 'to establish a multi-partner, blended tertiary education and training institution with on-campus facilities in Alice Springs and a network of remote study centres to serve non-urban desert communities'.[99] The idea of a shared educational campus had first been mooted during the

1990s. CAT and Batchelor signed a memorandum of understanding in 1999. The chairmen of both organisations, Bray and Gatjil Djerrkura, co-chaired a joint meeting of the DPC consortium at the Araluen Arts Centre in June 2000, which the chairman of the Institute of Aboriginal Development (IAD) also attended, expressing strong support for the project. IAD's board subsequently agreed in principle to join the consortium. The DPC steering committee met the Northern Territory cabinet later in the year and won the Chief Minister's backing, and then travelled to Canberra to lobby for federal support. However, the apparent consensus unravelled over the next two years, with CAT and Batchelor expressing frustration that IAD was not seriously committing itself to a shared campus, and CAT's board resolving that the organisation should withdraw from the project if funding guarantees were not received from the Territory and federal governments by the end of 2002. It was IAD, however, that formally withdrew from the DPC in December 2002.

By 2003 the DPC seemed stalled. CAT's board declared in May that the project had reached 'a low point and that they would like to rekindle enthusiasm'.[100] Planning and lobbying resumed, and in February 2004 the Northern Territory government announced major funding for the DPC, together with an ambitious timeline for the new Alice Springs campus to open in 2006, 'linking remote communities through a network of services and training that responds to the aspirations of people from within their cultural context'.[101] Some $10.4 million would be invested in the project by the Northern Territory and another $8.4 million by the Commonwealth. Walker called the DPC 'the single largest investment of capital to support Indigenous culture and education into the future in Australia'.[102] Construction began in 2004, although the key building work did not start until 2007. Behind the scenes, CAT's board sought to ensure that mainstream support did not dilute the partners' original purpose, and expressed

> concern … about a potential loss of focus on delivery of services and opportunity for desert people, who have growing and needy younger generations. For that reason, the Board confirms their understanding that the DPC is to be a catalyst for change in the desert and should maintain its resolve to support livelihoods for Indigenous people in desert Australia'.[103]

As planning and then construction proceeded, however, Bray expressed increasing confidence that 'an opportunity will arise through the Desert Peoples Centre to move [the consortium's Indigenous] training program national'.[104] CAT's growth over the period of planning for the DPC meant that by the time CAT was ready to move there was insufficient space available in the new

Federal Education Minister The Hon Julia Gillard opened the Desert Peoples Centre on 28 May 2010. Pictured: Peter Taylor (CAT CEO), Adrian Mitchell (BIITE Director), Dr Bruce Walker, Hon Julia Gillard MHR, Harold Furber (Chair DPC), James Bray (Chair CAT), Karl Hampton MLA (obscured), and Hon Warren Snowdon MHR.

buildings on the precinct to accommodate the entire organisation. With support from the Indigenous Land Corporation CAT purchased the adjacent 37 hectare site occupied by the CSIRO, refitted the buildings and connected the two sites with a new road. Bray's hard work was rewarded when in March 2010 all of CAT's operations moved from Priest Street to the new precinct, and in June 2010 the DPC was officially opened by Deputy Prime Minister Julia Gillard.

The relocation to the DPC seemed to symbolise a new beginning for CAT. It reaffirmed CAT's core principles and provided a stronger foundation for delivering them into the future. As Bray explained, 'at Priest Street we're landlocked. You're restricted in your vision and your dreams. We needed land to build on, to expand, to give our people a better education. We've made a step into the future by starting the Desert Peoples Centre'.[105] It was a bold step, but in full context an uncertain one. As Walker admitted during 2007, 'whilst we have tried hard at CAT over the years, we were not really in a position to cause lasting change because we are not powerful enough to turn heads. We have only been able to work within the conventional wisdom and space provided by Government policy and programmes'.[106]

3. Investing in the Outback

Despite, and indeed partly because of, the difficult policy environment within which CAT operated during the first decade of the 21st century, the organisation articulated an explicit and holistic strategy that was designed to secure the regional future of Outback Australia. In part, CAT's proposals were developed in response to the ascendancy of neoliberal thinking, which was reflected in the mainstreaming of education and training, the channelling of jobs and services into hub towns at the expense of smaller communities, the winding back of CDEP, the Northern Territory Intervention and the National Indigenous Reform Agreement of 2009. However, in a deeper sense, CAT's reform agenda expressed the board's independent vision of 'Happy and Safe Communities of Indigenous People' first expressed in *Planning for the Future* in 2003. It rested on the lessons learned from its community development activities, and its frustrations with government policy inconsistencies and failures over the previous 30 years. CAT's reform agenda was also informed by, and in turn reinforced, the expansion of the Desert Knowledge Consortium, which became Desert Knowledge Australia in 2001 and was relaunched as a statutory corporation in 2003. Walker had helped to launch the consortium and became a director in the new corporation, under the chairmanship of Fred Chaney (a board member of Reconciliation Australia and former federal Liberal parliamentarian, serving as Minister for Aboriginal Affairs in the Fraser government during the late 1970s). Its Desert Knowledge Precinct in Alice Springs, funded by the Northern Territory and Australian governments, expanded to accommodate the Desert Knowledge CRC, CSIRO and the DPC.

A central element in CAT's reform agenda was the DPC, which was

> developed from the vision of Aboriginal leaders and educators to address the failure of current education and training systems to develop the skills and capacities of Aboriginal people. Its vision is to establish a systematic and coordinated approach to education in order to foster a framework for future cultural, social and economic development.[107]

Importantly, the DPC linked education and development. Bray and Rose Kunoth-Monks (CAT board member and Chancellor of Batchelor) insisted that the DPC must play 'a key role in developing a thriving desert economy. The gathering and sharing of knowledge, strengthening human capital, and acting as a responsible and informed advocate of desert Australia and its peoples, is basic to this goal'.[108] However, the DPC was only the most visible expression of broader strategic thinking within CAT during the decade. CAT explicitly and systematically critiqued government policy interventions in Outback Australia and proposed a bold alternative strategy for sustainable community futures on a region-wide basis. CAT's board members had asked themselves at the time of ATSIC's abolition in 2003, 'what is the way forward for CAT given that we have always attempted to be one step ahead of the knee jerk response to issues that arise?'.[109] As the decade proceeded the way forward seemed increasingly clear. CAT contended that the debates raging within the capital cities about the 'presumed big issues of native title, governance, substance abuse, [and] family violence all serve to mask an underlying problem'. That problem was the absence of a comprehensive 'development approach or livelihoods model that drives the delivery of services and settlement development' throughout the Outback.[110] Moreover, neither development nor livelihoods could be sustained on the basis of government handouts alone. Global financial crises disclosed the limits of prudent government investment in national development. Neoliberalism made equally clear what would and would not be supported within such prudent national economic planning. As CAT acknowledged in 2008, 'there are not too many settlements around the world that survive and thrive when they are not associated with a significant economic resource or market adjacent to them'.[111]

CAT's concession did not mean that it had capitulated to neoliberal demands for the livelihoods of all Australians to be tied to the marketplace dictates of the formal economy or for the winding back of homeland communities that could not thrive within the national economy. CAT had always denied that the homelands movement represented a retreat from the modern world and it continued to do so after Vanstone ridiculed outstations as cultural museums. In 2007, for example, when Hughes argued in *Lands of Shame* that the homelands' appeal lay in enabling Aboriginal people to 'live traditional lives as hunter-gatherers uncontaminated by modern Australia', Walker returned from study leave in Spain to point out that in Central Australia 'there are many people who have retained culture and language but are able to access a 21st century lifestyle that engages with the world'.[112] Walker realised that in Spain culture and lifestyle were underpinned by a broad spectrum of work activities in both the formal and informal economies, opportunities that were not available in remote Australia. As Fisher noted in 2006, 'Around three-quarters of Indigenous people want to work in their own community, but there are not enough jobs. Around eighty per cent of existing positions are taken up by outsiders'.[113] Far from addressing this problem, homelands policy during the decade exacerbated it. CAT warned against the consequences of forcing smaller Aboriginal communities to seek jobs and services in larger urban centres, on the grounds that

> It is difficult to see how employment opportunities will be found should people abandon their communities and move to town. The social trauma and dysfunction currently faced by people is unlikely to be relieved by creating a series of fringe settlements or new suburbs around Darwin, Alice Springs, Cairns, Kalgoorlie, Mt Isa, Broome and Kununurra.[114]

Moran (who rejoined CAT in 2005) agreed that 'moving people from the settlements to regional towns, "where the jobs are", will leave the economic circumstances of people unchanged and may worsen their health and well-being'.[115]

CAT's continuing advocacy of Aboriginal Australians' right to live and work within a hierarchy of Outback towns and settlements did, however, acknowledge consequences and responsibilities as well as entitlements. CAT reaffirmed its proposition, clearly stated since the 1980s, that 'You don't live in the bush to experience city life or access services that only become viable if significant numbers of people use them … As a result, you don't solve problems in the bush by defining them in terms of services available in a city'.[116] The right to live in the bush did not carry with it the right to live in the bush as one would in the city. Appropriate technology could maximise livelihood options in the bush, but there nonetheless remained trade-offs to be made. CAT argued that government practice had obscured that reality:

> Over the past 30 years the rights espoused in international covenants and the race discrimination act and principles of equity for all Australians provided the policy rationale for service provision. The application of these covenants and principles has inherently been directed towards achieving a position of similarity or comparability. As a result efforts are directed at eliminating difference and disadvantage.[117]

CAT sought instead to achieve difference without disadvantage.

Many influential voices simultaneously questioned the thrust of government Aboriginal affairs policy, and claimed that dysfunction in Aboriginal settlements highlighted the 'massive failure' of the 'progressive approach based on rights' launched by Coombs during the brief reformist government of Gough Whitlam in the early 1970s and sustained into the new century by governments from both the left and right of federal politics.[118] CAT did not join these criticisms of its old friend, and pointed out that the goal of 'statistical equality' that it criticised in government policy owed more to the 'practical reconciliation and mainstreaming' of the Hawke, Keating and Howard years than to Whitlam.[119] It was the effect of policy change during those later years that CAT's board had in mind when it declared in 2003 that the negative 'effects of two decades of welfare dependency and a "rights" only agenda is apparent and needs to be turned around'.[120] Moran spelled out the board's concern, contending that a 'safety net' of human rights entitlements that did not also provide satisfying and useful activities for people could well 'lead to a "cargo cult" type of passivity'.[121] Much of Moran's experiences working with CAT had been as head of its Cairns office in north Queensland, and indeed CAT's position on regional community development overlapped with that being simultaneously articulated in north Queensland by Noel Pearson's Cape York Institute and by the Balkanu Cape York Development Corporation alongside whom CAT's Cairns staff worked. Pearson gained national attention in 2000 for his trenchant criticism of 'the "gammon" economy of passive welfare', within which the 'economy of [Aboriginal] communities is artificially sustained by government transfers' while their internal social relationships disintegrated because 'responsibility and reciprocity are [no longer] demanded in the transactions and relationships of society'.[122]

CAT's purpose across central Australia, like Balkanu's in Cape York, was to nurture the energies of the communities that were being stifled by this 'gammon' economy. Drawing attention to the 'practical limitations of a rights and equity based agenda', the board argued that 'Until non Indigenous people know what the financial cost/benefit of encouraging and supporting Indigenous people to practice culture and pursue livelihoods on country is[,] they will always work within the welfare dependency, supply driven developmental model that contributes to the current malaise'.[123] CAT's early move to establish Desert Services Australia in 2002 was intended as a means not only of delivering and commercialising its Bushlight services, but also of addressing the lack of employment options in Aboriginal communities by making more systematic use of the CDEP funds that underpinned the economies of many such settlements. During the decade CAT sought to harness local CDEP funding more fully so that community members could participate in installing and maintaining CAT technologies and running local services. A majority of CAT's training programmes were undertaken using CDEP funding. In doing this, and in the face of government efforts to wind back the operations of CDEP after the abolition of ATSIC, CAT thought deeply about broader livelihood options for places in which no conventional economy existed. Moran, for example, drew upon his international community development experience to insist that whereas mainstream Australia was underpinned by a market economy, as indeed were most communities supported by aid agencies in the 'developing world', in Outback Australia 'the economy of most remote Aboriginal settlements is dominated by government transfers'.[124] Moran contended that

> In place of markets in remote Indigenous settlements, local economies are dominated by public services, which include housing, water, telephones, power, roads, rubbish, health, education, banking, police, justice, aged care, sports, unemployment, child protection, and income support. Public services are variously administered by multiple agencies operating across three levels of government (Federal, State/Territory, and local), each with separate administrative requirements.[125]

CAT sought to rework and rearrange these dynamics so as to facilitate local decision making over the allocation of government resources and ongoing employment by local people. CAT also placed attention upon the potential benefits of encouraging what (with the exception of its partner Rio Tinto) was so conspicuously lacking in the Outback, that is, private enterprise engagement with Aboriginal communities, and also private enterprise initiatives by Aboriginal people. CAT's thinking was spelled out in two important policy papers during 2008. The first paper dealt with CDEP, the second with the Intervention.

Neoliberal analysts condemned CDEP for underpinning the supposedly false economies of homelands settlements by providing 'sheltered employment that would not expose Aborigines and Torres Strait Islanders to mainstream competition'.[126] In contrast, CAT saw in CDEP a means of revitalising Outback community economies if it could be agreed that the goal of 'sustainable Indigenous communities does not necessarily mean that the communities "earn a living" by individuals having paid employment in the generally accepted Western sense'.[127] CAT's review of CDEP operations in the Northern Territory, written by Taylor and Young, was completed in April 2008. It pinpointed the inadequacies of the existing system, arguing that implicitly the main purpose of CDEP as it was then administered was to subsidise local-government service delivery to Aboriginal communities rather than provide lasting employment pathways to their residents, and that at best CDEP delivered only low-skill jobs that did not translate into regular employment. Taylor and Young also drew attention to the broader context of entrenched Aboriginal disadvantage in education and training, the flow-on effects upon workforce participation, and the erosion of Aboriginal engagement in local decision making as community councils were absorbed into larger shires. CAT's report proposed reforms to improve skills acquisition and transition from CDEP into regular employment, together with better alignments between communities and potential employers in the mining and construction industries, land management programmes, art centres and tourism businesses, as well as more streamlined and accessible local government operations. CAT urged that reforms to CDEP should aim at 'revitalising local economies through a mix of enterprise, mainstream employment, education and community services, and building sustainable communities'.[128]

CAT's intention was to break open the conventional framework within which CDEP operated in order to free it from neoliberal rule making about work and development so that CDEP could be more readily applied to a much wider range of livelihood activities important to communities. As Moran argued, conventional economic thinking saw 'employment as the only valid form of productive activity. Yet there is a range of important productive activity engaged in by those who are unemployed'.[129] Walker, likewise, realised from his period in Spain

> the hundreds of small opportunities that people take to earn a living. At CAT we talk about enterprise and trading but tend to see it as an industry thing. [In Spain] I watched huge numbers of buskers who were lined up in all sorts of places entertaining with all manner of acts and poses. They had developed ways of getting people to pay and be photographed with them. There was skill in what they did and they appeared to make a reasonable income from that skill.[130]

Walker joked that rethinking notions of productive activity 'brought a whole new dimension to selling paintings in the [Alice Springs] mall'.[131] However, this rethink required people to break free from the assumption that traditional culture isolated people from the modern world and real-life economics. Walker worried that criticisms of the homelands movement obscured this important point, with the result that the Aboriginal 'settlement pattern driven by cultural attachment [to traditional lands] is now deemed problematic because Indigenous culture is seen to not support increased engagement in the mainstream and in the view of some has helped create the social dysfunction evident in many communities'.[132]

CAT used appropriate technology to interweave culture, place and livelihoods in the modern world. As Walker put it, appropriate technology could play 'a key part [in] determining and securing a livelihood and participating in the economy'.[133] Its use empowered people with aspirations to live on country without forgoing all the benefits of participating in the wider nation. As Walker remarked, 'CAT has always held that technology and science hold the seeds for indigenous development and economic independence.'[134] Balkanu argued likewise from its experience of the appropriate technologies being used on Cape York that the 'most conspicuous characteristic of successful outstation people is that they have a high degree of control over their lives. They are powerful people, especially when they are within their outstation domains'.[135] With those small domains coming increasingly under question in the early 21st century, CAT pragmatically emphasised the need for the aspirations and resources of local communities to be fitted into broader regional contexts and developmental opportunities. Walker increasingly turned his mind to the 'need to spend more time examining what might form the economic base for the dispersed network of settlements across [Outback] Australia'.[136]

CAT's submission to the Northern Territory Emergency Response Review Board in August 2008 captured this vision.[137] It was entitled *Investing in the Outback: A Framework for Indigenous Development within Australia*. The report ('auspiced' by CAT, but carefully worded in order to safeguard CAT by not necessarily 'representing' its views) was written by Walker in collaboration with Doug Porter (ATSIC reviewer of CAT during the 1990s, and subsequently an adviser to the World Bank) and Mark Stafford Smith (a principal scientist for the CSIRO in Alice Springs, co-convenor with Walker of the Desert Knowledge Consortium, and foundation CEO of the Desert Knowledge Cooperative Research Centre). The three men urged that policy debate must move beyond the specifics of child sexual abuse, for which targeted actions could be put in place with across-the-board sanctions; beyond the broader and more contentious rhetoric about welfare dependency, dysfunctional communities and failed states, which generated only circular momentum and partisan interest; and beyond the short-term paternalistic expediency of enforcing responsible family income management and collective community behaviour. Policy makers needed to address the long-term political economy of the overall region.

Investing in the Outback proposed that for policy debate to move forward along these lines, two aspects of current thinking had to change. Firstly, government funding should adopt a forward-looking focus on 'enabling livelihoods' in place of the prevailing backward-focused preoccupation with 'minimising disadvantage'.[138] Walker, Porter and Stafford Smith contended that the 'provision of services based on filling gaps or catching up might be desirable or equitable from a national perspective, but such a starting point should only be driven by an overwhelming philosophy of investment rather than deficit reduction. Continued reliance on welfare will have results that fall short of national expectations'.[139] The report reasoned that investment designed to maximise livelihood options in Outback communities would need to:

> encompass some local service roles such as mechanics, building construction and maintenance, and hairdressers; local government and community services such as road maintenance, settlement management and rubbish; services to society at large such as controlling weeds and fire, quarantine and caring for endangered species; opportunistic engagement where available with bigger business such as mines and tourism; all linked to a generous interpretation of cultural activities including art, music and performance, sport, but also traditional cultural activities on country that sustain language and other cultural capital that is valued by the nation.[140]

Secondly, the report advocated that state interventions should aim to enable economic livelihoods across the entire Outback, thus 'broadening the basis of the [Northern Territory Emergency] response to one of national interest rather than Indigenous interest and disadvantage'.[141] The futures of small Aboriginal communities (and of organisations such as CAT that supported them) would thus be linked with the larger dynamics of regional development in the national interest. The report proposed that this might be underwritten by an Australian Outback Trust Fund, and applied by a representative National Australian Outback Commission so as to ensure that the investment strategy was influenced by the local inputs of Aboriginal people. CAT's CDEP paper had similarly proposed the creation of two Regional Employment and Community Development Authorities (one of them Desert and the other Tropics) to coordinate the CDEP in partnership with government, employers and other agencies. *Investing in the Outback* also overlapped with interpretations and proposals that were being advocated by Desert Knowledge Australia. Its remoteFOCUS project, launched in 2008, resonated especially with CAT's thinking, which was not unexpected given that the project was led by Walker.

CAT's plan offered a mechanism whereby state interventions could 'assist Indigenous opportunity by guaranteeing investment in a pattern of settlement and services that enables the market to expand into the regions of need, and at the same time provide safety, security and service for all Australians in the national interest'.[142] Public investment could thus be harnessed to foster open-ended regional economic development by encouraging a diversity of local initiatives, rather than dictating options. Thus could 'the national interest and the interest of indigenous people be served through a vibrant network of settlements and people dispersed across our most remote regions' and sustained by a rich diversity of livelihood activities.[143] These activities would amount to what Jon Altman calls a 'hybrid economy', and provide tangible 'development alternatives for people living culturally and geographically beyond the mainstream'.[144] However, the Australian government's Indigenous Economic Development Strategy for 2011–18, released late in 2011, appears designed 'to sideline the potential of hybrid economic activities … in remote locations'.[145] The opportunity for fundamental change in the Outback has been delayed, but still exists. However, first of all, in Walker's words, 'As a nation Australia must fundamentally re-conceive and reconstruct its view of its own backyard and must understand the systems that sustain that fragile part of the nation known as the Outback'.[146]

Chapter 5

Conclusion

In 2010 CAT celebrated its 30th birthday. There was much to be proud of. CAT's alternative interventions in Outback Australia are generally acknowledged to have been sustained and successful. Yet many who work for CAT today are apprehensive about the organisation's future and the broader direction of Australian Aboriginal affairs. Some express misgivings that CAT has grown so big that it has become difficult to 'stay grounded' in Aboriginal communities, carrying with it the risk that 'we lose sight of where we are going and what we're here for in the first place'. Some speculate that as CAT has grown large its cutting edge has blunted: 'it was a place where you could break new ground … I don't think they're breaking new ground anymore, they're just treading on the same old ground: up and down, up and down'.

Some express nostalgia for CAT's early days when the Priest Street workshop was in full swing, turning out drum ovens and other homeland products, and first earning a reputation for CAT 'way out in the bush'.

Some are concerned that only a third of CAT's 120 staff are Aboriginal: 'we get good Aboriginal people in … we teach them … but we can't seem to keep them here so that we build up a pool'.

Some worry about the consequences of CAT's move to the DPC, whose new campus of attractive landscaped buildings cannot mask the enormity of the future challenges facing CAT as a 'catalyst for change' in Outback Australia.

Two concerns in particular are widely shared by CAT staff and board members. Firstly, they express dismay about the negative stereotypes of remote Aboriginal communities that circulate in mainstream Australia. They know first-hand the problems that do exist in many communities, but they also recognise the positives, and the diversity of outcomes. As CAT board member Dale Jones remarked,

> one of my main passions is informing people from the cities … what the actual circumstances are out in communities. They make judgements, and

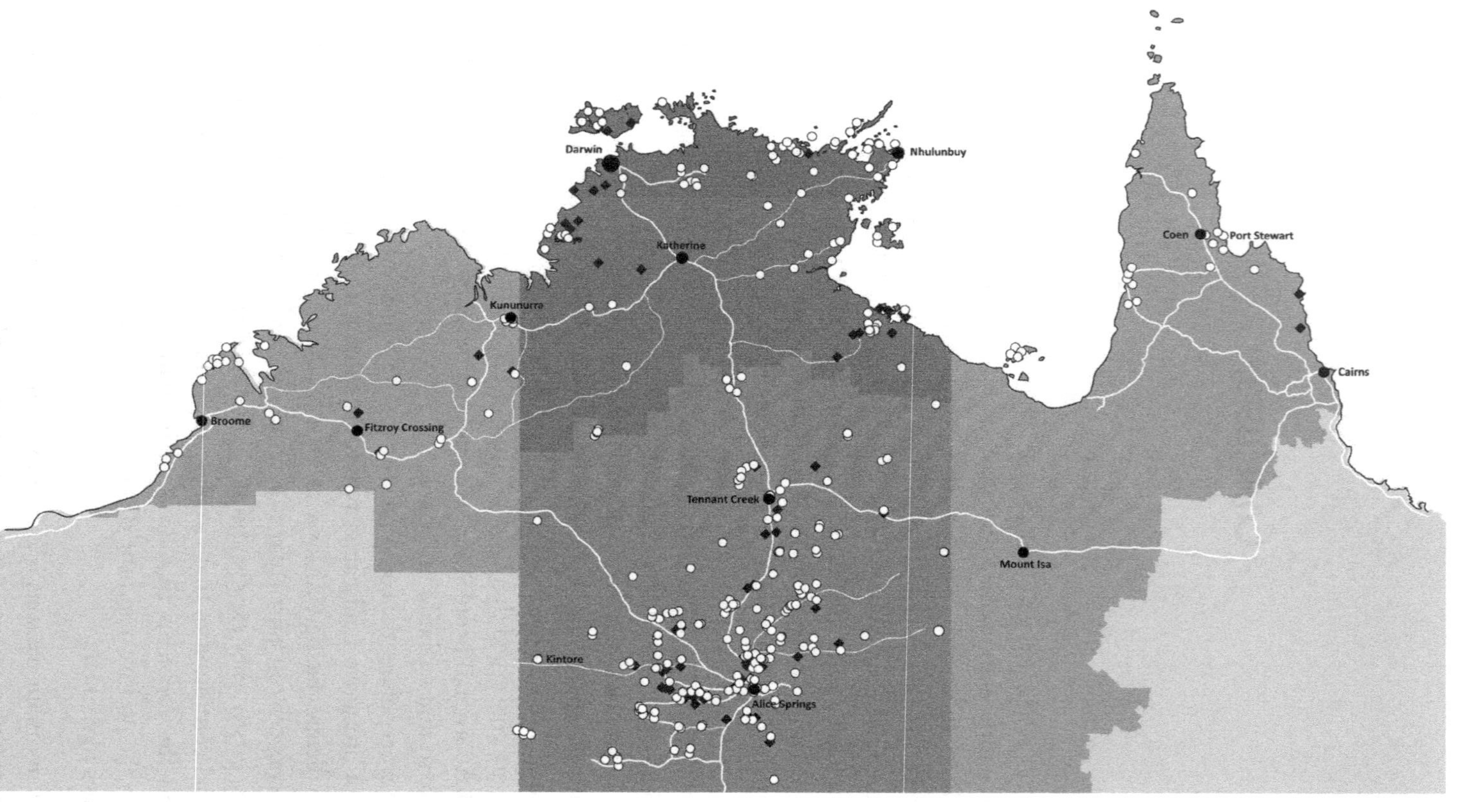

Communities who have participated in CAT's alternative interventions over the past 30 years.

> make assumptions, and most of them have never even set foot in [a remote community.] That can be very frustrating and very disheartening because they have no idea what they are talking about.[1]

Secondly, all express disappointment at the constraining effects of government policies and processes. They fear that participatory community development activities have disintegrated since CDEP was 'gutted' ('reformed', says the Australian government, with the caveat that CDEP 'is not intended to be a substitute for mainstream employment') and its funding redirected through the core agencies of the federal bureaucracy.[2] CAT board member Steve Hirvonen, who had been chairman of the ATSIC Gulf and West Queensland regional council, recalled that some $30 million in non-discretionary federal funds used to be channelled through the council annually, but that government departments had 'no idea' how best to allocate it and so left 'huge gaps in service delivery'. Hirvonen speculated, 'If I had that $30 million over three years, $90 million, and I invested it I reckon we could fund every social programme out in our region'.[3]

Although CAT takes pride in its history of alternative interventions, it has been largely dependent on government to fund those activities, and it is – as the experience of ATWORK highlights – vulnerable to the changing emphases of government policies. Such dependence is not absolute. Bruce Walker recalled that during CAT's first 15 years 'we were getting grants and we were setting the agenda; we would get an allocation and I'd decide what we'd spend it on',[4] and even as neoliberalism reshaped the policy and funding environment in more constraining ways, CAT's 'strategic alliance' with Peter Taylor, who was a senior federal bureaucrat in Canberra through the 1990s and into the mid 2000s, continued to provide key funding opportunities for CAT.[5] The origin of Bushlight exemplifies this productive relationship. Yet the ongoing effort required to sustain such initiatives in an increasingly difficult operating environment is unrelenting. Even Bushlight, despite the awards and recognition it has received, and the fundamental improvements to quality of life that it has achieved in remote communities, is vulnerable to changing government policy on remote communities. As Bushlight's former head, Grant Behrendorff, acknowledged,

> That's the challenge, I guess. During the homeland movement we were in the right place at the right time [but now] at a time when homelands lost favour … the grants follow the policies of the day … In CAT's case, there's not much money around to support research around people returning to homelands or trying to make homelands sustainable. They're only really

> interested about growth towns being sustainable, and homelands being basically about people spending weekends.[6]

Jim Bray stood down as Chairman in 2010, exhausted. He remarked, 'I'm losing faith in a lot of things. I'm tired … I'm fed up. We've been fighting for these things for a long time'.[7] CAT's long fight – against both the stereotypes and the unhelpful policy mindsets – poses an uncomfortable question to all Australians: is it Aboriginal and Torres Strait Islander society that is in deep crisis, or government policy and general community 'buy in' on Aboriginal affairs? Walker joined with Fred Chaney in 2011 to argue that the Australian government's Northern Territory Intervention in 2007 'was symptomatic of profound failures of governance' that had steadily accumulated since the mid 1990s.[8] Government interventions in Aboriginal affairs, intended to close the gap between Aboriginal and non-Aboriginal Australians, have paradoxically constrained the livelihoods of Aboriginal and Torres Strait Islander people. The basic premises behind such interventions are wrong: as Metta Young, one of CAT's senior managers, said, the policies are 'all [about]"lack" and "deficit", what you don't have and what you need to do, and then you go into mainstream employment driven by this underlying sense of trying to get people off their homelands or out of remote communities and into larger centres'.[9]

There are nonetheless important and positive lessons for the future that can be drawn from CAT's experiences in Outback Australia. There are good reasons for cautious optimism towards both CAT's continuing effectiveness and the possibility of broader advances in Aboriginal wellbeing. There is clear evidence of an overall improvement in the socioeconomic conditions of Aboriginal Australians over the last 35 years, notwithstanding the mixed results of government policy interventions.[10] It is imperative to build upon these gains, and not allow them to dissipate. Several points from CAT's history have special relevance in this regard.

The first point is that appropriate technology works in the Outback. Appropriate technology cannot be dismissed as a quaint footnote in history, interesting to Gandhians and those who remember the 1970s phrase 'small is beautiful', and used today only by a minority of 'green' environmental activists in some regions of the developed world. CAT's appropriate technology products like Bushlight systems and community phones, together with the research and design process that produces them, their regular servicing and the training and education programmes that inform their use, are making a difference in communities throughout remote Australia. Taylor was matter-of-fact when describing Bushlight as being 'widely regarded as one of the few long-term successes in remote Indigenous services'.[11] CAT's ongoing research and information sharing regarding water supply and water quality have also contributed significantly to Aboriginal wellbeing. These activities began with Walker's fieldwork at Lake Nash in 1981, were consolidated by the national water report for the federal Race Discrimination Commissioner in 1994, sustained by CAT's national research partnerships on water quality and treatment, and were most recently illustrated by CAT's 2009 field guide for community water management planning.

CAT is widely remembered for its first generation of appropriate technology products that were designed during the 1980s for long and reliable use on the homelands: the hand pumps, drum ovens, VIP toilets and the like. Their resilience is unquestioned: when Marc Siedel took over as manager of CAT's Western Australian operations in 2001, he found that in the Kimberleys 'VIP toilets were dotted all the way along the Great Northern Highway'.[12] However, some people sneered at CAT products as 'second-rate options' for Aboriginal people.[13] Ralph Folds questioned the appropriateness of 'appropriate technology' and argued that in remote locations such as Papunya and Walungurru the 'Pintupi revelled in using cars, televisions, refrigerators, air conditioners and video players and, like other Australians, many believe their lives are lacking or incomplete without them'. Walker's counter-argument was that 'You don't live in the bush to be like a city … You don't solve problems in the bush by defining them in terms of services available in a city'.[14] However, later generations of appropriate technology products and design approaches have made such exchanges obsolete. Brian Singleton, an Aboriginal project officer for CAT in Queensland, noted in 2010 that Bushlight systems in remote communities allowed problem-free use of 'little luxuries' such as fridges, fans, air conditioners, TVs and DVD players.[15] These fundamental lifestyle changes in the Outback make redundant the unhelpful suggestion from outsiders that the homelands movement represents a retreat by Aboriginal Australians into the Stone Age. They also make it possible to think in terms of 'a broader policy framework that goes beyond a simple trade-off – equality *or* difference – [to one that] is instead based on more coherent and inclusive notions of equality *and* difference'.[16] Warning her people against becoming 'locked into stationary orbit', Rose Kunoth-Monks argues that 'If we rely only on identifying with land and culture we become tied to customary practice rather than facing the future and what needs to be done today … No longer is reliance on land and culture sufficient'.[17]

A second point from CAT's past concerns constituencies. CAT has struggled to attract the sustained interest of policy makers in government, although it has won significant victories from time to time with both federal and state governments. However, CAT's prime concern has been to address the seemingly mundane issues of day-to-day living in local communities. CAT ATWORK trainer Robyn Ellis expressed this concern with local issues when she explained the social and physical considerations involved in designing something seemingly as simple and transmutable as a door handle. In Outback Australia,

> One of these houses [during a] Sorry Business time has 30 people using it. A lot of those are kids, lots of doors opening, closing, slamming, a lot of dust. Those little passage sets and door locks they put on those houses are actually designed for an urban environment where you have two adults and 2.5 children using it, who aren't there all day because they're at work or school. Those handles will work under those conditions, but you put them into this remote [place and] their grease barrels fill with dust, acts like grinding paste and it jams, the people get cranky, keys get lost, massive doors bang, bang, bang, you know, the door lock falls off, people don't have a screwdriver sitting there to tighten it up when it comes loose; all these factors make that the most inappropriate door lock you could put on, but they persist. In the 15 years I've been doing this job they keep putting the same locks on. Even though they know they're failing, no-one takes that step to say let's design something that won't fail.[18]

CAT's history demonstrates that local issues and local outcomes do count. Addressing them properly makes a real difference to local livelihoods and quality of life. Thus, appropriate technology in CAT's hands has been as much about listening to communities and giving them options as it has been about delivering door handles or solar energy systems.

A third point relates to community development. Taylor argued that substantial improvements in Aboriginal wellbeing will not be achieved until livelihood options for remote Aboriginal communities are redefined as 'a development issue'. Aboriginal wellbeing will not be achieved by dispensing well-meaning charity, but by helping communities to develop sustainable livelihoods. Australian foreign aid does this overseas. However, community development in an Australian context should not be limited to Australia's overseas aid programme, AusAID, and to community sponsorship of aid agencies overseas. It also needs to happen in Australia's own backyard. CAT's advocacy of appropriate technology has always been informed by international community development principles, and Taylor saw this as 'a really useful and productive way of engaging Indigenous people in discussion and debate about what they need, what they want, what they can afford, what is sustainable'.[19]

Back in the 1970s Jim Downing deplored the lack of attention given to community development principles in Outback Australia. In the eyes of Andre Grant, a CAT staffer in Cairns, things had not changed substantially by 2010: 'community development methodologies are not applied very much in Australia[;] for some reason Australia, because it's a developed nation not a developing nation, maybe ... decisions [are] made top-down from Canberra about what should happen in Aboriginal communities'.[20] However, CAT's participatory community development projects, especially since the pioneering work of Mark Moran in north Queensland during the 1990s, provide a blueprint for better approaches to Aboriginal community development from the ground up. CAT's initiatives were applied to specific communities, such as Old Mapoon and Port Stewart in north Queensland. However, CAT has increasingly complemented these tailored community development activities with broader advocacy of comprehensive development planning for Outback Australia as a whole. Walker has argued since the 1980s that without effective wealth creation activities in remote communities, the return to country could not break the dependence of Aboriginal people on government handouts.

Walker advocated passionately during the 1990s that service delivery must be driven by 'a national statement of Indigenous Peoples Rights'. However, as he prepared the water report he concluded that as a result of simplistic interpretations of human rights entitlements under the Racial Discrimination Act, the 'development of Aboriginal Australia (and the provision of technology and services) has been driven by the assumption that [Indigenous people] are disadvantaged, a deficit model.' This crude but well-intentioned approach will, he cautions, forever doom Aboriginal people to 'sitting at the back of the bus' and should be replaced by a forward-thinking regional development model which refocuses attention 'on investment in opportunities and [Indigenous] capacities rather than on problems and needs'.[21] Walker lobbied governments tirelessly about the need to develop a coherent planning approach for the Outback that harnessed its economic potential rather than simply delivering benefits and services, and since retiring as CAT's CEO, he has continued this advocacy role through Desert Knowledge Australia's remoteFOCUS project.[22] Walker and his colleagues insist upon 'the importance of economic policy rather than "services" as the key policy objective in remote communities'.[23] It is surely possible to support the emergence of a 'hybrid' regional economy that offers 'development alternatives for people living culturally and geographically beyond the mainstream'.[24]

Community development as it has been applied by CAT is an entirely different form of intervention to that practised by the Australian state (although the COAG-endorsed 'Closing the Gap' initiative commits all governments 'to rebuild relationships with Indigenous people').[25] CAT works in Aboriginal communities only upon their invitation and for as long as it is welcome there. Its activities are designed to meet the specific requests of the communities it partners with and to align with their cultural emphases. CAT's alternative interventions, developed especially through ATWORK and the national clearinghouse, thus entail a two-way flow of knowledge. CAT water specialist Robyn Grey-Gardner encountered the conventional one-way flow in the early 2000s when she participated in the CRC for Water Quality and Treatment as it sought to implement the findings of Walker's water report. The CRC assumed that improving water access for Aboriginal communities would be a simple matter of technology transfer, but 'it became apparent very quickly that it wasn't just a matter of opening doors for the wealth of expertise that was sitting at the CRC for Water Quality and Treatment, based in Sydney, Melbourne, and Adelaide'.[26] CAT has learned to employ a different approach to problem-solving. As Moran explained,

> I think that the good ideas in Indigenous affairs come from … relationships that are formed between trusted outsiders and Indigenous leaders. There's a new knowledge that generates there. It's … not traditional knowledge or necessarily local knowledge either. So, community doesn't necessarily understand or know what to do. I mean, it certainly [should be necessary to] have the locals in power when it comes to setting priorities, and identifying and defining the problems, but the solutions don't necessarily happen because the 'zone' is intercultural, so you actually need different types of knowledge coming together. So it's when you get someone from the outside going in, hanging around long enough, an Indigenous leader is prepared to take a step halfway to meet them, and they get on and listen to each other. That's when you get this new knowledge, and in a way I think that happened through Bruce's early work when he was going into communities, and then his relationship with Jim.[27]

A fourth key point from CAT's past is the benefit of accountable public service. CAT is a not-for-profit NGO that exists to serve the Outback rather than the political constituencies of mainstream Australia. There is an important role for such organisations in Aboriginal affairs because there is a great difference between providing services, thus locking in Aboriginal dependence, and serving communities by assisting them to develop resilience and wellbeing. Walker often challenged colleagues to ask, 'Does what I am doing reduce the dependence of the remote community on outside resources and does the activity enhance the quality of life of the individuals who live in the remote community?'[28] Such thinking epitomises public service and resonates also with Australians' long tradition of volunteerism for the public good. It is no accident that Walker and many of those involved in CAT's early years were active members of the Uniting Church, nor that Bray (along with other Aboriginal boys from Alice Springs who later became national leaders in their respective fields) was educated at the Anglican Church's St Francis House in Adelaide. Similarly, it is no accident that volunteerism has been an important and continuing element in CAT's activities. One of the ingredients in the board's success has been its insistence upon non-payment for members; Noel Bowen, who successfully managed the Priest Street workshop in its early years, initially worked as a volunteer; the installation of his workshop products in the homelands required Aboriginal volunteers; the Rio Tinto fellowship scheme depended upon its staff volunteering to work with CAT; Andre Grant's community development projects in north Queensland since 2008 have been supported by voluntary work contributed by Engineers Without Borders and university engineering students. CAT has always attracted staff who want to give more than they receive: 'bright-eyed and bushy-tailed', laughed one.[29]

After celebrating its 30th year in 2010, CAT began the second decade of the 21st century with a dynamic new CEO, Peter Taylor. His sudden death in 2013 is both a blow for CAT and a setback for Aboriginal advocacy in Australia. However, CAT has a fresh chairman, Peter Renehan, and a bold new five-year strategic plan. Its webpage masthead proclaims the important continuities in CAT's long-running Outback activities: 'Community Aspirations + Appropriate Technology = Sustainable Livelihoods'.[30] It is a potent and constructive message. As the board maintains, CAT 'doesn't exist purely because someone thinks it is a good idea. It has meaning and exists if people want it and it can satisfy their needs'.[31] It exists, in Jenny Kroker's words, 'to make a difference out there for our people'.[32]

Notes

Chapter 1 – Introduction

1 See Nicholas Biddle and Mandy Yap, *Demographic and Socioeconomic Outcomes across the Indigenous Australian Lifecourse: Evidence from the 2006 Census*, Australian National University E Press, Canberra, research monograph No. 31, 2010.

2 Bruce Walker, 'The Emperor's New Clothes', presentation at the Northern Territory Indigenous Housing Workshop, 11–12 July 2006.

3 Peter Taylor, recorded interview, 19 August 2008.

4 Bruce Walker, recorded interview, 19 June 2008.

5 Tim Rowse, *Indigenous Futures: Choice and Development for Aboriginal and Islander Australia*, UNSW Press, Sydney, 2002, p. 236.

6 Noel Hayes, recorded interview, 21 May 2009. Jon Altman called Closing the Gap a 'top-down approach' in his *Beyond Closing the Gap: Valuing Diversity in Indigenous Australia*, Centre for Aboriginal Economic Policy Research, Australian National University, Canberra, 2009, Discussion Paper No. 54/2009, p. 1.

7 Mark Moran, 'What Job, Which House? Simple Solutions to Complex Problems in Indigenous Affairs', *Australian Review of Public Affairs*, March 2009, available at www.australianreview.net/digest/2009/03/moran.html.

8 Walker, 'Science and Technology for Remote Communities', keynote address, Science and Technology for Remote Communities Conference, Murdoch University, July 1988 (copy in CAT archives, box 2).

9 Bruce Walker, Douglas Porter, and Ian Marsh, *Fixing the Hole in Australia's Heartland: How Government Needs to Work in Remote Australia*, Desert Knowledge Australia/ remoteFOCUS, Alice Springs, 2012, p. 5.

10 ibid., p. 8. See Alan Mayne (ed.), *Beyond the Black Stump: Histories of Outback Australia*, Wakefield Press, Kent Town, 2008; Alan Mayne and Stephen Atkinson (eds), *Outside Country: Histories of Inland Australia*, Wakefield Press, Kent Town, 2011.

11 Tim Rowse, *Remote Possibilities: The Aboriginal Domain and the Administrative Imagination*, North Australia Research Unit, Australian National University, Darwin, 1992, p. 19.

12 E.F. Schumacher, *Small is Beautiful: Economics as if People Mattered*, Harper & Row, New York, 1973, p. 183.

13 Patrick Dodson, 'Whatever Happened to Reconciliation?' in (eds) Jon Altman and Melinda Hinkson, *Coercive Reconciliation: Stabilise, Normalise, Exit Aboriginal Australia*, Arena Publications Association, North Carlton, 2007, p. 22.

14 Rosalie Kunoth-Monks, Foreword in Amnesty International, *The Land Holds Us: Aboriginal People's Right to Traditional Homelands in the Northern Territory*, Amnesty International Australia, Broadway, NSW, 2011, p. 4.

15 See Adrian Smith, 'The Alternative Technology Movement: An Analysis of its Framing and Negotiation of Technology Development', *Human Ecology Review*, vol. 12, no. 2, 2005, pp. 106–19.

16 See www.cat.org.uk

17 See www.cse.org.uk/pages/about-us/our-history

18 See 'The History of NCAT', at www.ncat.org

19 Dale Jones, recorded interview, 21 May 2009.

20 M.K. Gandhi, *Constructive Programme: Its Meaning and Place*, Naviajivan Publishing House, Ahmedabad, 1991, pp. 14, 15.

21 Schumacher, *Small is Beautiful*, pp. 20, 138–51.

22 ibid., pp. 174, 206.

23 ibid., p. 19.

24 ibid., p. 145.

25 Ban Ki-Moon, Foreword, in *The Millennium Development Goals Report, 2009*, United Nations, New York, 2009, p. 3.

26 See Porter's critique of liberal developmentalism in David Craig and Doug Porter, *Development Beyond Neoliberalism? Governance, Poverty Reduction and Political Economy*, Routledge, London, 2006.

27 For an early critique, see John Altman and Mike Dillon, 'Funding Aboriginal Development in Central Australia', in (eds) Barney Foran and Bruce Walker, *Science and Technology for Aboriginal Development*, CSIRO and CAT, Melbourne, 1986, np.

28 Mark Moran, recorded interview, 12 June 2010.

29 *CAT Plan 2003–2006: Planning for the Future*, Centre for Appropriate Technology, Alice Springs, nd, p. 4.

30 Quotations sourced from Walker's original 1985 TAGAL workshop notes in CAT archives, box 2, and their later publication in Bruce Walker, 'Integrated Development Planning', in Foran and Walker, *Science and Technology for Aboriginal Development*.

31 Bruce Walker, Editorial, *Our Place*, no. 16, 2001, p. 1.

32 Examples of localised innovations and appropriations are scattered through Walker's papers and presentations.

33 Rick Callahan, Report on Haasts Bluff, September 1984, CAT archives, box 13.

34 CAT, *Annual Report 1996*, Centre for Appropriate Technology, Alice Springs, p. 7.

35 *Our Place*, vol. 1, no. 2, June 1996, p. 4.

36 Mark Moran, recorded interview, 12 June 2010.

37 This wording is repeated in many CAT documents; see for example *Submission to the Northern Territory Government on the Review of the Community Development Employment Program*, Centre for Appropriate Technology, Alice Springs, April 2008, p. 2.

38 CAT Plan 2003–2006, p. 4.

39 See J.C. Altman, *In Search of an Outstations Policy for Indigenous Australians*, Centre for Aboriginal Economic Policy Research, Australian National University, Canberra, 2006, Working Paper 34/2006.

40 CAT, *Review of the National Homelands Policy*, Stage 1 (Draft), August 2004, p. 15.

41 Foreword to D.L. Japanangka and P. Nathan, ,Kibble Books & Central Australian Aboriginal Congress, Malmsbury, Victoria, 1983, p. vii. See Tim Rowse, *Obliged to be Difficult: Nugget Coombs' Legacy in Indigenous Affairs*, Cambridge University Press, Cambridge, 2000.

42 J.C. Altman, S. Kerins, B. Fogarty and K. Webb, Why the Northern Territory Government Needs to Support Outstations/Homelands in the Aboriginal, Northern Territory and National Interest, Centre for Aboriginal Economic Policy Research, Australian National University, Canberra, 2008, Topical Issue Paper No. 17/2008, p. 2.

43 Brian Singleton, recorded interview, 24 May 2010.

44 Quoted in Moran, 'What Job, Which House?'. See also Sean Kerins, *The First-Ever Northern Territory Homelands/Outstation Policy*, Centre for Aboriginal Economic Policy Research, Australian National University, Canberra, 2008, Topical Issue Paper No. 09/2009.

45 Walker, 'Sustainability in Indigenous Australia', conference presentation, Murdoch University, 12–14 July 2006, p. 5. See J.C. Altman and L. Taylor, , Australian Government Publishing Service, Canberra, 1987.

46 See the transcript of interview from ABC Radio's 'AM' programme, 9 December 2005, at www.abc.net.au

47 Jim Downing, *Ngurra Walytja: Country of my Spirit*, North Australia Research Unit, Australian National University, Darwin, 1988, p. 158. Japanangka and Nathan, *Settle Down Country*, p. 4.

48 Ralph Folds, *Crossed Purposes: The Pintupi and Australia's Indigenous Policy*, University of New South Wales Press, Sydney, 2001, p. 121.

49 Science & Technology for Aboriginals in Outstation Communities: A Report of the Science & Technology Workshop held in Central Australia, 22–25 September, 1980, DAA, February 1981, CAT archives, box 2.

50 Rowse, *Indigenous Futures*, p. 5.

51 Walker, recorded interview, 19 June 2008.

52 Walker, 'Study Leave Report, March–May 2007', CAT Board Meeting, 7 February 2007.

53 Jenny Kroker, recorded interview, 22 August 2008.

Chapter 2 – Beginnings: the 1980s

1 CCCA had until recently been the Alice Springs Community College, and later became Alice Springs College of TAFE (Technical and Further Education), then Centralian College. It is now part of Charles Darwin University.

2 Walker to Jon Cook, 7 December 1984, in CAT archives, box 5.

3 Walker to Northern Territory Department of Education, 23 June 1989, in CAT archives, Water Report series, box 1/20.

4 Tim Rowse, *Remote Possibilities: The Aboriginal Domain and the Administrative Imagination*, North Australia Research Unit, Australian National University, Darwin, 1982, p. 99.

5 Charles Perkins, *Welfare and Aboriginal People in Australia – Time for a New Direction*, University of New England, Armidale, NSW, 1990, p. 9. See Tim Rowse, *Indigenous Futures: Choice and Development for Aboriginal and Islander Australia*, UNSW Press, Sydney, 2002.

6 Notes from an interview with Jim Bray, 28 January 2009.

7 Rowse, *Remote Possibilities*, p. 19.

8 *Australian*, 26–27 May 1979, 'Darwin Community College', in Bruce Walker personal archives.

9 ibid.

10 Basil Hetzel, recorded interview, 9 December 2009.

11 Walker, recorded interview, 24 August 2008.

12 Walker to H.C. Coombs, 16 April 1986, CAT archives, box 5.

13 See Tim Rowse's sympathetic study of Coombs, *Obliged to be Difficult: Nugget Coombs' Legacy in Indigenous Affairs*, Cambridge University Press, Cambridge, 2000.

14 Jim Downing, *The Myth of Self Determination: Evidence given to the Senate Standing Committee on the Social Environment*, Institute for Aboriginal Development, Alice Springs, 1975, p. 2.

15 Walker, 'Technology and Employment in Aboriginal Communities: An Appropriate Training Option', at the APACE Second E.F. Schumacher Memorial Conference on Technology and Employment, 27–29 November 1986, Morphet, NSW, in CAT Archives, box 1.

16 See Rowse, *Obliged to be Difficult*, pp. 141–42.

17 Basil Hetzel, Malcolm Dobbin, Lorna Lippmann, Elizabeth Eggleston (eds), *Better Health For Aborigines? Report of a National Seminar at Monash University*, University of Queensland Press, St Lucia, Qld, 1974, p. 72.

18 ibid., pp. xix, 243–56. See Downing, *The Myth of Self Determination*, pp. 12–13.

19 Basil Hetzel, recorded interview, 9 December 2009.

20 Neville Perkins, 'The Program of the Central Australian Aboriginal Congress in Relation to Nutrition and Agriculture', in B.S. Hetzel and H.J. Frith (eds), *The Nutrition of Aborigines in Relation to the Ecosystem of Central Australia. Papers presented at a Symposium CSIRO 23–26 October 1976 Canberra*, CSIRO, Melbourne, 1978, p. 11.

21 ibid., p. 1.

22 ibid. Note Hetzel's reaction against the 'rubbishing of Coombs in recent years' in Basil Hetzel, recorded interview, 9 December 2009.

23 See the preface by Hetzel in Hetzel and Frith, *The Nutrition of Aborigines*, pp. 1–2.

24 Downing, *The Myth of Self Determination*, p. 2.

25 ibid., p. 1. See Ross Howie, 'Obituaries: Jim Downing 1926–2009', *Australian Aboriginal Studies*, no. 1, 2009, p. 125.

26 G.F. Griffin and C. Lendon, *A Report on Visits Through Three Aboriginal Homelands in Central Australia*, CSIRO Division of Land Resources Management, Perth, November 1979.

27 Walker, recorded interview, 8 August 2008.

28 B.R. Ede to DAA, 4 March 1980, CAT archives, box 9.

29 Jim Pearse to Walker, January 2010, in Walker's personal papers.

30 ibid.

31 ibid.

32 Ian Viner was replaced as Minister for Aboriginal Affairs in December 1978 by Fred Chaney.

33 Pearse to Walker, January 2010.

34 B.G. Hansen to CCCA, 22 February 1980, and attachment, 'Aboriginal Health. Alice Springs, 17–18 September 1979', in CAT archives, box 9.

35 ibid.

36 'Lecturer in Appropriate Technology', undated typescript in Walker's personal papers.

37 Pearse took a one-year fellowship at the University of New England, and was thereafter redeployed into the senior executive service of the Territory public service in Darwin. He then moved to the Darwin Institute of Technology (formerly DCC), which in 1989 became part of the new Northern Territory University (re-established as Charles Darwin University in 2004).

38 G.W. Chard to Commonwealth Department of Education Alice Springs, 22 May 1980, in Walker's personal papers.

39 G.W. Chard to Walker, 20 March 1980, ibid.

40 ibid.

41 Notes by Walker, 3 June 1980, in ibid.

42 DAA, 'Some anticipated outcomes for Friday 20/6/80', manuscript in CAT archives, box 9.

43 'Notes of the meeting on the role of Appropriate Technology held in the conference room at D.A.A. Alice Springs at 2 p.m. on Friday, 20th June, 1980', typescript in Walker's personal papers.

44 H. Tinsley to all lecturing staff, 16 July 1980, 'TAFE Regional Co-Ordinator', ibid.

45 H. Tinsley to Walker, 'Responsibilities and Procedures', 4 September 1980, ibid.

46 Walker to H. Tinsley, 'Responsibilities and Procedures', 5 September 1980, ibid.

47 G. Chard to H. Tinsley, 17 September 1980, ibid.

48 G. Chard to H. Tinsley, no date [September 1980], ibid.

49 H. Tinsley to Walker, 5 November 1980, 10 March 1981, ibid.

50 H. Tinsley to Walker, 10 March 1981, 'Bi-Monthly Report for January and February, and 1981 Programme', in ibid.

51 Walker to H. Tinsley, 12 March 1981, '1981 Programme', ibid.

52 B. Cruise to All Staff, 'N.T.T.S. Organizational Structure', 4 May 1982, ibid.

53 Walker, recorded interview, 4 February 2009.

54 Walker, manuscript notes on memorandum from Principal to Heads of School, 'Aboriginal Training', 15 October 1982, in Walker's personal papers.

55 Walker, manuscript notes on meeting with M. Ivory and G. Marks, 21 May 1980, ibid.

56 Walker, recorded interview, 4 February 2009. See DAA, *Science and Technology for Aboriginals in Outstation Communities. A Report of the Science and Technology Workshop held in Central Australia, 22–25 September, 1980*, (DAA, February 1981), in CAT archives, box 2.

57 Walker subsequently assisted the United Nations Capital Development Fund in the Shan States of Myanmar in 1991 and 1992, and the United Nations Drug Control Program in 1998. In 2000 he was invited to speak at an APEC Forum in Mexico by invitation from the Japanese government, and in China at a Village Power Workshop in 2002 at the invitation of the United States government.

58 Downing, *The Myth of Self Determination*, p. 12.

59 Walker, 'TAGAL 1980–1985', at 'The Application of Science and Technology to Aboriginal Development in Central Australia: A Workshop', 8–10 October 1985, typescript in ibid., box 1.

60 J. Downing, 'The Lake Nash Saga – And Its Solution', February 1980, typescript in ibid., box 13.

61 Walker, 'Trip to New Bore (Mt Leibig), 29 July 1980', manuscript in Papunya Outstations folder, ibid., box 13.

62 Walker, recorded interviews, 24 August 2008, 4 February 2009. See Geoffrey Bardon, *Papunya Tula: Art of the Western Desert*, McPhee Gribble, Ringwood, Vic, 1991; Geoffrey Bardon and James Bardon, *Papunya: A Place Made After the Story*, Miegunyah Press, Melbourne, 2004.

63 Sam Fellows, 'Centre invents "bush-tech"', *Digest of Australia's Northern Territory*, vol. 10, no. 4, December 1988, p. 35.

64 Walker to DAA Darwin, 19 February 1981, in CAT archives, box 5.

65 Northern Territory Department of Education to Walker, 2 April 1984, ibid., box 3. Walker, Submission for the Establishment of an Appropriate Technology Extension Service, 30 April 1984, in CAT archives, unnumbered archives box (general correspondence, 1984–2004).

66 Minutes of Aboriginal Communities Draft Funding Programme Workshop, 27 April 1984, ibid., box 9.

67 Sam Fellows, 'Centre invents "bush-tech"', *Digest of Australia's Northern Territory*, p. 35.

68 Walker to all staff, 29 April 1985, in CAT archives, box 12.

69 Walker to Kurt Seemann, 27 August 1986, ibid., box 5.
70 Metta Young, personal communication, 28 May 2013.
71 Bruce Walker, personal communication, 8 July 2013.
72 Walker to CEP Secretariat, Darwin, 17 May 1984, ibid., box 3.
73 Walker to Docker River Council, 28 May 1984, ibid., box 13.
74 CAT, *Appropriate Technology in Central Australia*, occasional paper no. 2, August 1984, ibid., box 10.
75 Kurt Seemann, personal communication, 19 March 2013.
76 'A.T. Workshop/ Workshop Summary' [held Alice Springs 7–9 April 1987], typescript in CAT archives, unnumbered archives box.
77 Recorded interviews: Walker, 19 June 2008, 4 February 2009; Bray, 19 October 2009.
78 Recorded interviews: Jim Bray, 19 October 2009; also Bray, 28 January 2009, Walker, 4 February 2009.
79 Bray, recorded interview, 19 October 2009; Kurt Seemann, personal communication, 20 March 2013.
80 Minutes of ASCOT Executive Council Meeting, 20 July 1987, p. 3, in CAT archives, box 16.
81 Metta Young, personal communication, 28 May 2013.
82 'Agenda Notes: Meeting to discuss future organisation of Centre for Appropriate Technology', 24 August 1984, ibid.
83 Minutes of Special Council Meeting, 14 October 1986, ibid.
84 'Agenda Notes: Meeting to discuss future organisation of Centre for Appropriate Technology', 24 August 1984, ibid.
85 Jim Downing, *Ngurra Walytja: Country of my Spirit*, North Australia Research Unit, Australian National University, Darwin, 1988, p. 145.
86 Walker to Dr Richard Lim, Chairman of Council, Alice Springs College of TAFE, 12 October 1987, in CAT archives, box 5.
87 Minutes of Special Council Meeting, 14 October 1986, ibid., box 16.
88 Secretary of Department of Education to the Minister for Education, 15 June 1988, 'Centre of Appropriate Technology', together with memo by Minister Tom Harris, 28 June 1988, ibid., unnumbered archives box (Jim Bray, Board meetings, miscellaneous).
89 See CAT Briefing Paper 24/11/88, in ibid.; also Minutes of Alice Springs TAFE Executive Committee, 20 July 1987, in box 5.
90 Walker to B. Ede, 17 May 1981, ibid..
91 TAGAL Workshop papers, 8–10 October 1985, in ibid., box 1; and 'Report to the Technical Advisory Group to Aboriginal Homeland Communities (T.A.G.A.L.). September 1980–September 1982', ibid., box 2.
92 Minutes, Interdepartmental Water Needs Committee, 26 March 1981, 28 May 1981, ibid., box 9.
93 Terry Smith to Walker, 19 June 1984, ibid., box 5.
94 Walker, speaking notes, TAGAL workshop, October 1985, ibid., box 2.

95 Walker, concluding address, Science and Technology for Remote Communities Conference, 18–19 July 1988, Murdoch University, ibid.

96 Walker, 'Is Housing Appropriate for Aboriginal People?', presentation at Second World Congress of Building Officials, 21–26 October 1990, Darling Harbour, in ibid., unnumbered archives box. Walker drew upon comments by Clive Scollay earlier in the year; see 'Walungurru Thoughts/ Clive Scollay', undated typescript [1990], ibid., unnumbered archives box (Matthew Parnell papers).

97 Walker, speaking notes, TAGAL workshop, October 1985; and Walker, keynote presentation, 'Science and Technology for Remote Communities', 1988, both in ibid., box 2.

98 Walker, 'Science and Technology for Remote Communities'.', ibid.

99 Walker, 'Appropriate Technology at the Community College of Central Australia. A Preliminary Statement', 18 June 1980, typescript in ibid., box 5.

100 Walker, 'The role of appropriate technology', manuscript notes for a presentation at an In-Service Training Course, TAFE Southern Region, 28–31 October 1980, ibid., box 3. Walker's words jar with those of Indigenous Affairs Minister Amanda Vanstone 15 years later in her infamous 'cultural museums' speech (see chapter 4).

101 DAA, 'Science and Technology for Aboriginals in Outstation Communities. A Report of the Science and Technology Workshop held in Central Australia, 22–25 September, 1980', ibid., box 2.

102 Walker to Alison Worrell, Regional Director of Community Nursing, Kimberley Region, Derby, 12 December 1988, ibid., box 6. See Downing's defence of CAT in *Ngurra Walytja*, p. 140, and Ralph Folds' later criticisms in *Crossed Purposes: The Pintupi and Australia's Indigenous Policy*, University of NSW Press, Sydney, 2001 p. 165.

103 'Trip to Mantamaru (Jamison), 18–29 September 1983', manuscript, Cat archives, box 13.

104 Clive Scollay, 'Walungurru Thoughts' [1990], ibid., unnumbered archives box marked 'Matthew Parnell'.

105 Walker to Gary Young, Community Adviser Haarts Bluff, 15 June 1984, ibid., box 5.

106 Atitjere Aboriginal Community, 27 September 1984, ibid.

107 Kaltukatjara Community Council to Walker, 23 June 1985, ibid.

108 Waringarri Aboriginal Corporation to Walker, 4 November 1988, ibid., box 6.

109 Kurt Seemann to Rainy Limanto, 5 April 1990, ibid.

110 Walker to W. Baird, 4 September 1989, ibid.

111 Walker, 'The role of appropriate technology', manuscript notes for a presentation at an In-Service Training Course, TAFE Southern Region, 28–31 October 1980, ibid., box 3.

112 ibid.

113 Bruce Walker and Barney Foran, 'Progress in Appropriate Technology and Land Management since the 1980 Workshop', in Barney Foran and Bruce Walker (eds), *Science and Technology for Aboriginal Development*, CSIRO and CAT, Melbourne, 1986, np.

114 Walker, keynote presentation, 'Science and Technology for Remote Communities', 1988, CAT archives, box 2.

115 Walker to National Aboriginal and Islander Health Organisation, nd [1989], ibid., 1994 Water Report, box 1/20.

116 Walker, 'Integrated Development Planning', presentation to TAGAL's 'The Application of Science and Technology to Aboriginal Development in Central Australia: A Workshop', 8–10 October 1985, ibid., box 1; later published as 'Integrated Development Planning' in Foran and Walker, *Science and Technology for Aboriginal Development*, np.

117 Walker, 'Aboriginal and Islander Water Supplies: The Five R's', presentation to the 'Arid Zone Water: A Finite Resource' conference, 11–14 April 1991, Alice Springs, ibid., box 2.

Chapter 3 – Realignments: the 1990s

1 David Craig and Doug Porter, *Development Beyond Neoliberalism? Governance, Poverty Reduction and Political Economy*, Routledge, London, 2006, p. 11.

2 H.C. Coombs, October 1991, quoted in Tim Rowse, *Obliged to be Difficult: Nugget Coombs' Legacy in Indigenous Affairs*, Cambridge University Press, Cambridge, 2000, p. 195.

3 'Notes of Meetings at Kintore Re CAT Design Document', 26 November 1997, typescript in CAT archives, unnumbered archives box (administration, Walungurru, water).

4 ATSIC Memo, 26 May 1996, ibid.

5 Peter Sutton, *The Politics of Suffering: Indigenous Australia and the End of the Liberal Consensus*, second edition, Melbourne University Press, Carlton, 2011, p. 58.

6 Kurt Seemann, personal communication, 20 March 2013.

7 Walker, recorded interview, 4 February 2009.

8 National Advisory Committee meeting, 22–23 May 1997, in Register of CAT Board Minutes, 1996–2007.

9 Andrew Lane, recorded interview, 24 May 2010.

10 Seemann, personal communication, 20 March 2013.

11 'In Twenty Years Time: Where Will You Be – How Relevant Will You Be. The Future of Some Aboriginal Organisations in Alice Springs', undated typescript in Walker's personal papers, p. 3.

12 Greg Phillips, 'View', 22 September 1995, typescript, CAT archives, unnumbered file labelled 'National Technical Services Clearinghouse. Reports'.

13 Mark Moran and Bruce Walker, recorded interview, 12 June 2010.

14 Seemann, personal communication, 20 March 2013.

15 Walker, recorded interview, 8 August 2008.

16 Walker to Northern Territory Department of Education, 23 June 1989, CAT archives, 1994 Water Report, box 1/20.

17 'Questions Related to CAT', undated (c. June 1992) 3 page typescript, CAT archives, unnumbered archives box (Jim Bray miscellaneous).

18 Dr Richard Lim to Walker, 24 July 1992, ibid.
19 Minutes of Executive Committee of Council of ASCOT, 4 August 1992, ibid.
20 Walker to the Secretary of the College Council, 30 June 1992, attachment in Dr Richard Lim to Walker, 24 July 1992, ibid.
21 'CAT Inc Staff Meeting', 24 June 1992, typescript in ibid.
22 ibid.
23 CAT Business Committee meeting, 7 September 1993, ibid.
24 Aboriginal Programs Employment and Training Advisory Council, minutes of meeting, 27–28 October 1993, ibid.
25 CAT board meeting, 7 September 1993, in CAT archives, unnumbered archives box (board meetings 1993–98).
26 Seemann, personal communication, 20 March 2013.
27 'Questions Related to CAT', nd (c. June 1992) 3 page typescript, CAT archives, unnumbered archives box (Jim Bray miscellaneous).
28 'CAT Inc Staff Meeting', 24 June 1992, typescript in ibid.
29 CAT board meeting, 14 October 1992, ibid.
30 Recorded interviews: Walker, 24 August 2008; Jim Bray, 19 October 2009.
31 'CAT Organisation Review: Queries, Puzzles and Proposals Group. Summary of Group Discussions During November 1995', CAT archives, unnumbered file labelled 'National Technical Services Clearinghouse. Reports'.
32 ibid.; Greg Phillips, 'View', 22 September 1995, ibid.
33 CAT board meeting, 29 March 1996, in CAT archives, unnumbered archives box (board meetings 1993–98).
34 Doug Porter and Russell Fisher, *National Technology Resource Centre, Centre for Appropriate Technology Inc. Evaluation Report*, September 1997, pp. 1–2.
35 Walker, 'In Twenty Years Time', p. 2.
36 ibid., pp. 1, 3.
37 CAT Board Paper, 'Better Services Through A Stronger Organisation', December 1997, p. 7, CAT archives, box 73.
38 Porter and Fisher, *National Technology Resource Centre*, p. 37.
39 ibid., p. 2.
40 CAT 'Better Services Through A Stronger Organisation', p. 6.
41 Draft minutes, National Advisory Committee, 24–25 November 1999, CAT archives, unnumbered archives box (board meetings, miscellaneous, 1996–2002).
42 Walker, 'In Twenty Years Time', p. 4.
43 Whiteboard printout from CAT workshop and board meeting, 13–15 December 1999, Alice Springs Resort, CAT archives, unnumbered archives box (board meetings, miscellaneous, 1996–2002).
44 Minutes from the previous board workshop in Alice Springs, 13–15 December 1999, included in the board papers for a two-day workshop at Glen Helen, 27–29 July 2001, ibid., unnumbered archives box (board meetings, 1996–2002).

45 Walker to Kroker, 20 December 1996, ibid., unnumbered archives box (Jenny Kroker: Women's Technology Network papers, 1996).

46 *ATWORK Newsletter*, no. 1, August 1992; ibid., unnumbered archives box (board meetings, miscellaneous, 1996–2002).

47 Chairperson's Report, in CAT, *Annual Report*, 1997, p. 2.

48 Sonja Peter and Christian Tietz, field trip to Docker River, 15–18 May 1995, CAT archives, box 71 (Docker River folder).

49 Walker, recorded interview, 19 June 2008.

50 'Opening of WCC Residential Complex and Premises. Smithy Zimran', nd (1991) 2 page typescript in CAT archives, unnumbered archives box (Matthew Parnell, miscellaneous).

51 Letter by Richard Callahan to ATSIC, 2 February 1994, ibid., unnumbered archive box (Matthew Parnell: CAT projects, 1992–95).

52 Undated 3 page ms minute by Walker to Kurt Seemann and Matthew Parnell, attached to a letter by the WCC Administrator Peter Johnsen to Seemann, 10 July 1995, ibid.

53 Jim Sinatra and Phin Murphy, 'Social Factors Affecting Planning and Development in Kintore, NT', RMIT Outreach Australia Program, 1997, ibid., unnumbered archive box (Joy Love: Kintore, 1995–97).

54 Mark Moran, recorded interview, 12 June 2010.

55 CAT board meeting, 18 March 1995, CAT archives, unnumbered archives box (board meetings 1993–98).

56 CAT board meeting, 6 September 1997, in Register of Board Minutes, 1996–2007.

57 ibid.

58 CAT board meeting, 16 July 1998, CAT archives, unnumbered archives box (board meetings, 1996–2002).

59 Walker to Robert Genders, CAT Development Engineer, 4 May 1995, 2 page ms, 'RE: feedback as requested on your Kintore Report', ibid., unnumbered archive box (Matthew Parnell: CAT projects, 1992–95).

60 Quoted in Ralph Folds, *Crossed Purposes: The Pintupi and Australia's Indigenous Policy*, University of NSW Press, Sydney, 2001, p. 173.

61 Alan Randell, 'Ngurra Pala Kanyilpai: The proper management of our homeland. Personal report for the Walungurru/Kintore Design Process Team', preliminary draft, December 1997, CAT archives, unnumbered archive box (Joy Love: Kintore, 1995–97).

62 'Community Meeting', 28 January 1997, one-page typescript, WCC, ibid.

63 Bruce Walker and Matthew Parnell, 'Extract of Draft Review of Sanitation Case Study of the Walungurru Community, Kintore, May 1996', attachment to ATSIC Memo, 26 May 1996, ibid., unnumbered archive box (administration, Kintore).

64 Walker, Director's Report, 18 November 1998, ibid., unnumbered archives box (board meetings, 1996–2002).

65 WCC, 'Report of Lilliah McCulloch', 18 October 1998, ibid., unnumbered archives box (workshop, Walungurru Community Council, 1998–99).

66 CAT board meeting, 18 November 1998, ibid., unnumbered archives box (board meetings, miscellaneous, 1996–2002).

67 CAT, 'National Technology Resource Centre', December 1993, pp. 3, 6, ibid., unnumbered file labelled 'National Technical Services Clearinghouse. Reports'.

68 Porter and Fisher, *National Technology Resource Centre*, p. 4.

69 'What is the NTRC?', *Our Place*, vol. 1, no. 1, February 1996, p. 5.

70 Kurt Seemann, personal communication, 20 March 2013.

71 Porter and Fisher, *National Technology Resource Centre*, Main Report, p. 4; Executive Summary, p. i.

72 ibid., p. 32.

73 ibid., Executive Summary, p. iv. CAT, 'National Technology Resource Centre', December 1993, p. 2, CAT archives, unnumbered file labelled 'National Technical Services Clearinghouse. Reports'.

74 'What is the NTRC?', *Our Place*, vol. 1, no. 1, February 1996, p. 4.

75 'Summary of Proceedings of a Discussion with Doug Porter (A Prospective Evaluator For CAT's NTRC Project)', 3 page typescript, undated, CAT archives, unnumbered file labelled 'National Technical Services Clearinghouse. Reports'.

76 Porter and Fisher, *National Technology Resource Centre*, Executive Summary, p. i.

77 National Advisory Committee meeting, 22–23 May 1997, in Register of CAT Board Minutes, 1996–2007.

78 Porter and Fisher, *National Technology Resource Centre*, Executive Summary, p. i.

79 Memorandum by Ron Talbot to Walker, 2 November 1997, 'Feedback concerning NTRC Evaluation Report (30 October)', CAT archives, box 73.

80 Su Groome, recorded interview, 25 May 2010.

81 Kurt Seemann, personal communication, 20 March 2013.

82 Health Habitat was established by Yami Lester in 1985.

83 CAT board meeting, 15 December 1999, in Register of CAT Board Minutes, 1996–2007. The energy forum had been held in Alice Springs in September 1999.

84 Barry Deane, 'A Report on the Potential for Collaboration between RTZ/CRA and the Centre for Appropriate Technology', 30 September 1996, in CAT archives, unnumbered archives box (Rio Tinto: Fellowships, Conferences, Partnerships). See Paul Wand, recorded interview, 13 June 2010.

85 Draft minutes, CAT-Rio Tinto meeting, 30 March 1999, in CAT archives, unnumbered archives box (Rio Tinto: Fellowships, Conferences, Partnerships).

86 *Our Place*, edition 12, January 2000, p. 15.

87 ibid., 'The Women's Technology Network', vol. 1, no. 1, February 1996, p. 6.

88 Robyn Ellis, recorded interview, 13 June 2010. See Jenny Kroker, recorded interview, 22 August 2008.

89 Robyn Ellis, recorded interview, 13 June 2010.

90 Ruth Hodson, 'Evaluation of the Aboriginal and Torres Strait Islander Women's Technology Network Conference', Centre for Aboriginal Studies, Curtin University, no date, in CAT archives, unnumbered archives box (Jenny Kroker: Women's Technology Network papers, 1996).

91 Women's Technology Network Conference, back cover advertisement in *Aboriginal and Islander Health Worker Journal*, vol. 20, no. 3, May/June 1996.

92 National Advisory Committee meeting, 22–23 May 1997, in Register of CAT Board Minutes, 1996–2007.

93 Allison Adams, 'Women with ATSIC: Technology Survey. Attitudes and Perceptions Concerning Technical Issues in Rural and Remote Indigenous Communities', CAT, August 2000; CAT archives, unnumbered archives box (Jenny Kroker: Women's Technology Network, 1996).

94 Walker, 'Technology and Employment in Aboriginal Communities: An Appropriate Training Option', presentation at the APACE Second E.F. Schumacher Memorial Conference on Technology and Employment, November 1986, Morpeth, NSW.

95 Kurt Seemann, personal communication, 19 March 2013.

96 Residential Report, in board meeting papers, 12 December 1997, Register of CAT Board Minutes, 1996–2007.

97 Russell Fisher, 'ATWORK: Curriculum Directions for the ATWORK Aboriginal Technical Worker Program', CAT, June 1995, p. 7, CAT archives, box 73.

98 Robyn Ellis, recorded interview, 13 June 2010.

99 'ATWORK Feasibility Study', undated draft report, CAT archives, box 71. See also the final booklet, *The Aboriginal Technical Worker Feasibility Study Report*, CAT, May 1990, ibid., box 74.

100 Fisher, 'ATWORK', p. 6.

101 See Kurt W. Seemann, 'Technacy Education: Understanding Cross-Cultural Technological Practice', in John Fien, Rupert Maclean, and Man-Gon Park (eds), *Work, Learning and Sustainable Development: Opportunities and Challenges*, Springer, Dordrecht, the Netherlands, 2009, pp. 117–31. Kurt Seemann, personal communication, 20 May 2013.

102 Seemann, in *Evidence before Senate Standing Committee on Employment, Education and Training*, 14 June 1991, p. 2429, CAT archives, unnumbered box (Jim Bray miscellaneous). See also Seemann, 'Introduction to Technacy: Concepts, Curriculum and the Technology Message Experienced in Remote Aboriginal Communities', paper at the First National Science and Technology Communicators Conference, Canberra, August 1990, ibid., box 74.

103 CAT, *Annual Report*, 1994, p. 5.

104 Kurt Seemann and Ron Talbot, 'Technacy: Towards a Holistic Understanding of Technology Teaching and Learning among Aboriginal Australians', *Prospects*, vol. 25, no. 4, December 1995, p. 773.

105 Robyn Ellis, recorded interview, 13 June 2010.

106 Ron Talbot, recorded interview, 15 June 2010.

107 Robyn Ellis, recorded interview, 13 June 2010; also Jenny Kroker, recorded interview, 20 May 2009.

108 Fisher, 'ATWORK', p. 5.

109 Ron Talbot, 'The ATWORK Programme: Is It A Representative Curriculum?', typescript, August 1992, CAT archives, box 71.
110 Walker, in *Evidence before Senate Standing Committee on Employment, Education and Training*, 14 June 1991, p. 2423.
111 Presentation by Seemann recorded in 'A.T. Workshop/ Workshop Summary', Alice Springs, 7–9 April 1987, typescript in CAT archives, black binder folder inside unnumbered archives box.
112 'ATWORK Report, 87/88', comments dated November 1987 in typescript in ibid.
113 Yirrkala, November 1987, in ibid.
114 Walker, 'Evidence before Senate Standing Committee on Employment, Education and Training', pp. 2424–25.
115 Ron Talbot, recorded interview, 15 June 2010.
116 Ron Talbot to Walker, 'Feedback concerning NTRC Evaluation Report (30 October)', memorandum dated 2 November 1997, CAT archives, box 73.
117 Untitled 25 page typescript (with handwritten attribution to Jonathan Dobbs, 1997), p. 12, ibid., unnumbered file labelled 'National Technical Services Clearinghouse. Reports'.
118 Recorded interview in Cairns, 24 May 2010.
119 Ron Talbot, recorded interview, 15 June 2010.
120 'ATWORK Report, 87/88', comments from Numbulwar, 10 November, 1987(?).
121 Director's Report, CAT board meeting, 14 June 1996, CAT archives, unnumbered archives box (board meetings, miscellaneous, 1996–2002).
122 Ron Talbot, recorded interview, 15 June 2010.
123 Walker, recorded interview, 24 August 2008.
124 Ron Talbot and Jenny Kroker, 'One Step Forward and Two Steps Back: The National Training Framework and Barriers to Indigenous Technical Training', presentation at the 1998 ATSIPTAC Conference 'Challenging Pathways', pp. 2–3, CAT archives, box 73.
125 Walker, recorded interview, 13 June 2010.
126 Greg Phillips, 'View', 22 September 1995, typescript, CAT archives, unnumbered file labelled 'National Technical Services Clearinghouse. Reports'.
127 Central Land Council to Irene Moss, 28 February 1990, ibid., 1994 Water Report, box 2/20.
128 'Launch of Water Report. Federal Race Discrimination Commissioner', 8 page typescript, 31 May 1994, ibid., box 1/20.
129 Irene Moss to Pat Dodson, 27 July 1988, ibid. Emphasis in the original.
130 'Launch of Water Report. Federal Race Discrimination Commissioner', ibid.
131 ibid.
132 Porter and Fisher, *National Technology Resource Centre*, p. 4.
133 ibid.
134 Walker, 'Evidence before Senate Standing Committee', p. 2427.
135 Folds, *Crossed Purposes*, p. 165.

136 Protest letter to Neil Bell MLA, 31 January 1991, copy attached to fax from Kurt Seemann to Smithy Zimran at Kintore, 22 March 1991, CAT archives, black binder folder in unnumbered box (Kintore/ Walungurru).

137 Seemann to Zimran, 22 March 1991, ibid.

138 'ATWORK Report, 87/88', comments from Galiwin'ku, 10 November 1987.

139 Robert Genders to Walker, 'Visit to Kintore on 26, 27th April 1995', 3 May 1995, CAT archives, unnumbered archive box (Matthew Parnell: CAT projects, 1992–95).

140 Walker, 'Aboriginal and Islander Water Supplies: The Five R's', presentation at the 'Arid Zone Water: A Finite Resource' conference, Alice Springs, April 1991, ibid., box 2.

141 *Newsletter of the CAT*, no. 1, Winter 1990, in file labelled 'Master Copies of CAT Products'. The words are repeated in CAT, *Annual Report*, 1995, p. 7.

142 CAT, *Annual Report*, 1996, p. 5.

143 Porter and Fisher, *National Technology Resource Centre*, p. 6.

144 Board meeting, 6 September 1997, in Register of CAT Board Minutes, 1996–2007.

145 Walker, recorded interview, 19 June 2008.

146 Walker, 'Aboriginal and Islander Water Supplies: The Five R's'.

147 'Walungurru Thoughts. Clive Scollay', 7 page undated typescript, 1990, in CAT archives, unnumbered archives box (Matthew Parnell, miscellaneous).

148 Walker, 'Is Housing Appropriate For Aboriginal People?' Presentation at the Second World Congress of Building Officials, Sydney, October 1990, ibid., unnumbered archives box (general correspondence, miscellaneous, 1984–2004).

149 Federal Race Discrimination Commissioner, *Water: A Report on the Provision of Water & Sanitation in Remote Aboriginal & Torres Strait Islander Communities*, AGPS, Canberra, 1994, p. 100.

150 Draft paper for CAT Board, 'Opportunities and Challenges for CAT', December 1999, 6 page typescript, CAT archives, unnumbered archives box (board meetings, miscellaneous, 1996–2002).

151 Walker, 'Evidence before Senate Standing Committee', p. 2432.

152 Federal Race Discrimination Commissioner, *Water,* pp. 163–64.

153 Tim Rowse, *Indigenous Futures: Choice and Development for Aboriginal and Islander Australia*, University of NSW Press, Sydney, 2002, p. 235.

154 Porter and Fisher, *National Technology Resource Centre*, p. 20.

155 Walker, recorded interview, 19 June 2008.

156 *Feasibility Report. Old Mapoon: Planning for a Healthy Community*, CAT, April 1994, p. 5; Jim Sinatra and Phin Murphy, *Planning for a Healthy Community: Old Mapoon*, Sinatra-Murphy, Melbourne, for Queensland Mapoon Aboriginal Community, Queensland Tropical Health Unit, CAT, 1995, p. 4.

157 ibid., p. x.

158 *Feasibility Report. Old Mapoon*, p. 32.

159 Mark Moran, recorded interview, 12 June 2010.

160 *Feasibility Report. Old Mapoon*, p. 19.

161 Draft paper for CAT Board, 'Opportunities and Challenges for CAT'.

162 'Organisational Review and Strategic Plan 1996–2000', in minutes of CAT board meeting 29 March 1996, CAT archives, unnumbered archives box ((board meetings 1993–98).

163 Porter and Fisher, *National Technology Resource Centre*, p. 5.

164 'A.T. Workshop/ Workshop Summary', Alice Springs, 7–9 April 1987, typescript in CAT archives, black binder folder inside unnumbered archives box.

165 ATWORK Report, 87/88, Galiwin'ku, 10 November, 1987.

166 Ibid., Milingimbi Island, 20 November, 1987.

167 *Feasibility Report. Old Mapoon*, p. 25.

168 Mark Moran and Gary Jones, *Planning Study. Doomadgee: Planning for Outstation Housing*, CAT, 1994, p. vi. C & B Consulting Group, *Community Settlement Plan: Old Doomadgee*, 2000, vol. 2 (Main Report), p. 4.

169 Draft paper for CAT Board, 'Opportunities and Challenges for CAT'.

170 Aboriginal Programs Employment and Training Advisory Council, minutes of meeting, 27–28 October 1993, CAT archives, unnumbered archives box (Jim Bray miscellaneous).

171 'Proposed Memorandum of Understanding', attachment to a fax sent to Djerrkura and Bray from the Institute for Aboriginal Development, 5 July 2000, ibid., unnumbered archives box (board papers, 2000–04).

Chapter 4 – Aboriginal Futures: the 2000s

1 Board meeting, 26 August 2002.

2 *CAT Plan 2003–2006: Planning for the Future*, Centre for Appropriate Technology, Alice Springs, nd, p. 4.

3 ibid., p. 3. Chairman's Report, in Minutes of AGM, 26 September 2003.

4 James Bray to Ian Trust (Chairman ATSIC WA State Council), 2 August 2002, CAT archives, unnumbered archives box (CAT board, 2000–04). Director's Report, in Minutes of AGM, 26 September 2001.

5 Board meeting, 28 February 2002, in Register of CAT Board Minutes, 1996–2007.

6 Board meeting, 26 August 2002.

7 Director's Report, ibid.

8 *Planning for the Future*, p. 5. See J.C. Altman, *In Search of an Outstations Policy for Indigenous Australians*, Centre for Aboriginal Economic Policy Research, Australian National University, Canberra, 2006, Working Paper 34/2006; Sean Kerins, *The First-Ever Northern Territory Homelands/Outstation Policy*, Centre for Aboriginal Economic Policy Research, Australian National University, Canberra, 2008, Topical Issue Paper No. 09/2009.

9 *Australian*, 22 May 2002, 'Hungry children feed our shame'.

10 *Planning for the Future*, pp. 6–7.

11 Steve Fisher, 'Discussion Paper on the Future Direction of the National Technical Services Clearinghouse of CAT' (Draft, 26 October 2001), in board meeting, 29 November 2001, CAT archives, unnumbered archives box (board meetings, 1996–2002).

12 Walker, recorded interview, 19 June 2008.

13 See Australian Government, *Closing the Gap on Indigenous Disadvantage: The Challenge for Australia*, Commonwealth of Australia, Canberra, February 2009; see also the annual *Closing the Gap: Prime Minister's Report,* for the years 2010–13, available at: www.fahcsia.gov.au.

14 Mark Moran, 'Neoliberalism and Indigenous Affairs', *Australian Review of Public Affairs* (April 2008), available at www.australianreview.net/digest/2008/04/moran.html.

15 Folds, *Crossed Purposes: The Pintupi and Australia's Indigenous Policy*, University of NSW Press, Sydney, 2001, p. 1.

16 Amanda Vanstone, 'Beyond Conspicuous Compassion: Indigenous Australians Deserve More Than Good Intentions', in John Wanna (ed.), *A Passion for Policy: Essays in Public Sector Reform*, ANU E Press, Canberra, 2007, pp. 40, 45.

17 *The Age*, 23 May 2006, 'Hate Stalks a Community where Gangs rule Roost'.

18 Patrick McCauley, 'Wadeye: Failed State as Cultural Triumph', *Quadrant Magazine*, vol. LIII, no. 12, December 2008.

19 Helen Hughes, *Lands of Shame: Aboriginal and Torres Strait Islander 'Homelands' in Transition*, Centre for Independent Studies, St Leonards, 2007.

20 Peter Sutton, *The Politics of Suffering: Indigenous Australia and the End of the Liberal Consensus*, Melbourne University Press, Carlton, 2009, 2011.

21 Walker, 'Sustainability in Indigenous Australia', presentation at Murdoch University, July 2006, p. 10, in Walker's personal papers.

22 Glen Helen workshop, 27–29 July 2001, in Register of CAT Board Minutes, 1996–2007.

23 Board meeting, 25 August 2003.

24 Board meeting, 26 August 2002.

25 Glen Helen workshop, 27–29 July 2001.

26 CAT, *Annual Report,* 2001–02, p. 4.

27 ibid., p. 7.

28 Board meeting, 26 August 2002.

29 CAT, *Annual Report*, 2001–02, pp. 4, 7.

30 Director's Report, ibid., p. 5. Glen Helen workshop, 28 July 2001. Director's Report, Minutes of 2002 AGM.

31 *Planning for the Future*, p. 4. A copy of the entire internal working document, 'CAT Plan 2003-2006: Planning for the Future' is filed in the office of the CAT Executive Officer.

32 'Our Place radio: a journey in the making', *Our Place*, no. 22, 2004, p. 12.

33 'CAT in Transition – A Structured Refocussing of the Organisation', 7 page typescript stamped 'Confidential', undated and unattributed, in CAT archives, unnumbered archives box (CAT board, 2000–04).

34 ibid.; also minutes of the Glen Helen board meeting and workshop, 14 July 2004, in Register of CAT Board Minutes, 1996–2007.

35 Board meeting, 26 August 2002.

36 Steve Fisher, 'A Review of the CAT Workshop', 26 February 2002, in minutes of board meeting, 28 February 2002, CAT archives, unnumbered archives box (board meetings, 1996–2002), and in papers for the same meeting, unnumbered archives box (CAT board, 2000–04).
37 Marc Seidel, recorded interview, 14 June 2010.
38 Robyn Grey-Gardner, recorded interview, 14 June 2010.
39 Peter Taylor, recorded interview, 19 August 2008.
40 ibid.
41 Board meeting, 13 March 2006.
42 Grant Behrendorff, recorded interview, 24 May 2010.
43 'CAT in Transition – A Structured Refocussing of the Organisation'; also minutes of the Glen Helen board meeting and workshop, 14 July 2004, in Register of CAT Board Minutes, 1996–2007.
44 Grant Behrendorff, recorded interview, 24 May 2010.
45 Director's Report, in board meeting, 26 August 2002, CAT archives, unnumbered archives box (CAT board, 2000–04).
46 Board meeting, 20–21 September 2007.
47 Board meeting, 7 April 2000, CAT archives, unnumbered archives box (CAT board, 2000–04).
48 'Succession Planning' workshop, 12 December 2002, Register of CAT Board Minutes, 1996–07.
49 CAT board workshop, 10–12 April 2003, ibid.
50 ibid.
51 Jim Bray, recorded interview, 8 August 2008.
52 Board meeting, 20–21 September 2007.
53 Fisher, 'Discussion Paper on the Future Direction of the National Technical Services Clearinghouse'.
54 Fisher, 'A Review of the CAT Workshop'.
55 Director's Report in CAT, *Annual Report*, 2001–02, p. 5.
56 Board meeting, 28 February 2002, CAT archives, unnumbered archives box (CAT board, 2000–04).
57 Board meeting, 26 August 2002.
58 'CAT in Transition – A Structured Refocussing of the Organisation'; also minutes of the Glen Helen board meeting and workshop, 14 July 2004, in Register of CAT Board Minutes, 1996–2007.
59 Fisher, 'Discussion Paper on the Future Direction of the National Technical Services Clearinghouse'. Emphasis in the original.
60 'G. Ramsey Notes', from the 12 December 2002 CAT board 'Succession Planning' workshop run by consultant Gregor Ramsay, 'Succession Planning' workshop, 12 December 2002, CAT archives, unnumbered archives box (CAT board, 2000–04).
61 'CAT in Transition – A Structured Refocussing of the Organisation'.
62 Board meeting, 13 March 2006.

63 Marc Seidel, recorded interview, 14 June 2010.

64 Walker, recorded interview, 4 February 2009.

65 Grant Behrendorff, recorded interview, 24 May 2010. ATSIS (Aboriginal and Torres Strait Islander Services) and DIMIA (Department of Immigration and Multicultural and Indigenous Affairs) were created by the Howard Government to replace ATSIC. DIMIA was replaced by the Department of Families, Community Services and Indigenous Affairs (FaCSIA), which subsequently became the Department of Families, Housing, Community Services and Indigenous Affairs (FaHCSIA).

66 Robyn Grey-Gardner, recorded interview, 14 June 2010.

67 Metta Young, recorded interview, 14 June 2010.

68 Board meeting, 26 August 2002, CAT archives, unnumbered archives box (CAT board, 2000–04).

69 'Report of Board Visit to NW Western Australia', board meeting, 27 November 2003, in Register of CAT Board Minutes, 1996–2007.

70 AGM Minutes, 27 September 2004.

71 COO Report, 29 June 2007.

72 Fisher, 'Discussion Paper on the Future Direction of the National Technical Services Clearinghouse'.

73 Paul Wand, recorded interview, 13 June 2010.

74 Glen Helen board meeting and workshop, 14 July 2004, in Register of CAT Board Minutes, 1996–2007.

75 Board meeting, 13 March 2006.

76 Minutes of AGM, 20 September 2007.

77 Papers from the Glen Helen workshop, 27–29 July 2001, including minutes from the workshop of 13–15 December 1999, CAT archives, unnumbered archives box (board meetings, 1996–2002). Walker had first framed these words in 'Opportunities and Challenges for CAT' (December 1999), ibid., unnumbered archives box (board meetings, miscellaneous, 1996–2002).

78 Fisher, 'Discussion Paper on the Future Direction of the National Technical Services Clearinghouse'.

79 Chairman's Report, in Minutes of AGM, 24 September 2002.

80 Board meeting, 13 March 2006, Register of CAT Board Minutes, 1996–2007.

81 Board meeting, 21 September 2007.

82 ITS report, 13 March 2006.

83 ibid.

84 Noel Hayes, recorded interview, 21 May 2009.

85 Walker, recorded interview, 12 June 2010.

86 Mark Moran, recorded interview, 12 June 2010.

87 COAG, National Partnership Agreement on Remote Service Delivery, 1 January 2009, p. 8, Schedule A: A-1. Available from the Standing Council on Federal Financial Relations: www.federalfinancialrelations.gov.au/content/npa/other/remote_service_delivery/national_partnership.pdf

88 Australian Government, *Closing the Gap on Indigenous Disadvantage: the Challenge for Australia*, February 2009, p. 6.

89 Northern Territory Government, *A Working Future*, 20 May 2009, available from www.workingfuture.nt.gov.au.

90 *Planning for the Future*, p. 7.

91 Metta Young, recorded interview, 14 June 2010.

92 Board meeting, 29 May 2002, CAT archives, unnumbered archives box ((CAT board, 2000–04).

93 CAT, *Annual Report*, 2001–02, p. 17.

94 Robyn Ellis, recorded interview, 13 June 2010.

95 Ron Talbot, recorded interview, 15 June 2010.

96 Chief Executive Officer's Report in CAT, *Annual Report*, 2007–08, p. 5.

97 Kurt Seemann, personal communication, 19 March 2013.

98 *Planning for the Future*, p. 21. The wording was widely repeated. See Bruce W. Walker, 'Building a Catalyst for Change in the Desert: The First 10 Years of the Desert Peoples Centre', presentation at the 2009 conference of the Council of Educational Facility Planners International, Australasia Region.

99 CAT, *Annual Report*, 2001–02, p. 24.

100 Board meeting, 2 May 2003, Register of CAT Board Minutes, 1996–2007.

101 Media release, 'Desert Peoples Centre Off and Running', 12 February 2004, CAT archives, unnumbered archives box (meetings, Bushlight).

102 Bruce W. Walker, 'Community, Art and Technology', presentation at the symposium 'Reclaiming the Centre: Art, Technology and Community in Outback South Australia and the Northern Territory', University of South Australia, September 2008, p. 1.

103 Board meeting, 17 March 2004, Register of CAT Board Minutes, 1996–2007.

104 Bray in Minutes of AGM, 20 September 2007.

105 Jim Bray, recorded interview by Glenn Morrison, October 2009, CAT.

106 "CEO Study Leave Report', March–May 2007, CAT board meetings, 2007.

107 *Review of the Community Development Employment Program: Submission to the Northern Territory Government*, CAT, Alice Springs, April 2008, p. 6.

108 James Bray and Rosalie Kunoth-Monks, 'To Dream the Impossible Dream: The Desert Peoples Centre', 5 page typescript in Bruce Walker's personal papers.

109 Board meeting, 25 August 2003.

110 ibid.

111 Bruce W. Walker, Doug J. Porter, and Mark Stafford Smith, *Investing in the Outback: A Framework of Indigenous Development within Australia. Submission to the Northern Territory Emergency Response Review Board*, CAT, Alice Springs, August 2008, p. 21. The paper was later published by the Academy of Social Sciences in its journal *Dialogue*, no. 28, February 2009, pp. 19–32.

112 Hughes, *Lands of Shame*, p. 11. 'CEO Study Leave Report' (2007).

113 *Our Place*, 'Livelihoods and services on country; achieving our goals', no.27, 2006, p. 7.

114 Walker, Porter, and Stafford Smith, *Investing in the Outback*, p. 21.

115 Mark Moran, 'What Job, Which House? Simple Solutions to Complex Problems in Indigenous Affairs', *Australian Review of Public Affairs* (March 2009), available at www.australianreview.net/digest/2009/03/moran.html.

116 Walker, Porter, and Stafford Smith, *Investing in the Outback*, p. 20.

117 ibid., p. 24.

118 Sutton, *The Politics of Suffering*, p. 7.

119 Walker, Porter, and Stafford Smith, *Investing in the Outback*, p. 22.

120 *Planning for the Future*, p. 7.

121 Moran, 'What Job, Which House?'

122 Noel Pearson, *Our Right to Take Responsibility*, Noel Pearson and Associates, Cairns, 2000, pp. 26, 30, 85.

123 Board meeting, 25 August 2003.

124 Moran, 'What Job, Which House?'

125 Moran, 'Neoliberalism and Indigenous Affairs'.

126 Hughes, *Lands of Shame*, p. 13.

127 Bray and Kunoth-Monks, 'To Dream the Impossible Dream'.

128 *Review of the Community Development Employment Program*, p. 2.

129 Moran, 'Neoliberalism and Indigenous Affairs'.

130 'CEO Study Leave Report', 2007.

131 ibid.

132 Walker, Porter, and Stafford Smith, *Investing in the Outback*, pp. 20–21; and Walker, 'Community, Art and Technology', p. 10.

133 ibid., p. 1.

134 'CEO Study Leave Report', 2007.

135 Balkanu: Cape York Development Corporation, *Cape York Outstation Report* (draft, prepared for the Peninsula Regional Council of ATSIC, August 2002; copy in CAT Cairns library), p. 101.

136 Walker, 'Community, Art and Technology', p. 10.

137 Walker, Porter, and Stafford Smith, *Investing in the Outback*.

138 Walker, Porter, and Stafford Smith, *Investing in the Outback*, pp. 27, 23.

139 ibid., p. 6.

140 ibid., p. 27.

141 ibid., p. 5.

142 ibid., p. 4.

143 Walker, 'Community, Art and Technology', p. 11.

144 Jon Altman, 'What Future for Remote Indigenous Australia? Economic Hybridity and the Neoliberal Turn', in Jon Altman and Melinda Hinkson (eds), *Culture Crisis: Anthropology and Politics in Aboriginal Australia*, University of New South Wales Press, Sydney, p. 270.

145 Ian Marsh, 'Are there Structural Barriers to Whole-of-Government?', in Fred Chaney, 'Submission to Senate Standing Committee on Community Affairs:

Stronger Futures in the Northern Territory Bill and the Social Security Legislation Amendment Bill 2011', RemoteFOCUS/ Desert Knowledge Australia, Alice Springs, 1 February 2011, Attachment A, p. 23.

146 Walker, Porter, and Stafford Smith, *Investing in the Outback*, p. 4; also Walker, 'Community, Art and Technology', p. 12. Walker had often made this point; see for example 'Speaking Notes. Desert Knowledge CRC: Chief Minister's Breakfast Meeting with Board members on 4 May 2004', 3 page typescript in Bruce Walker's personal papers.

Chapter 5 – Conclusion

1 Dale Jones, recorded interview, 21 May 2009.

2 John Hopkins, recorded interview, 24 May 2010. Australian Government, *Closing the Gap: Prime Minister's Report 2012*, Commonwealth of Australia, Canberra, February 2012, p. 91; Australian Government, *Closing the Gap: Prime Minister's Report 2013*, Commonwealth of Australia, Canberra, February 2013, p. 30.

3 Steve Hirvonen, recorded interview, 21 May 2009.

4 Bruce Walker, recorded interview, 13 June 2010.

5 Peter Taylor, recorded interview, 19 August 2008.

6 Grant Behrendorff, recorded interview, 24 May 2010.

7 Jim Bray, recorded interview, 13 June 2010.

8 Fred Chaney, Bruce Walker, Ian Marsh, Doug Porter, John Huigen, 'Submission to Senate Standing Committee on Community Affairs', Desert Knowledge Australia/ remoteFOCUS, Alice Springs, February 2011, p. 3.

9 Metta Young, recorded interview, 14 June 2010.

10 See Jon C. Altman, Nicholas Biddle and Boyd H. Hunter, *How Realistic are the Prospects for 'Closing the Gaps' in Socioeconomic Outcomes for Indigenous Australians?* Centre for Aboriginal Economic Policy Research, Australian National University, Canberra, 2008, Discussion Paper No. 287/2008. Republished as 'Prospects for 'Closing the Gap' In Socioeconomic Outcomes for Indigenous Australians?' *Australian Economic History Review*, vol. 49, no. 3, November 2009, pp. 225–51. See also *Closing the Gap: Prime Minister's Report 2013*.

11 Peter Taylor, recorded interview, 19 August 2008.

12 Marc Siedel, recorded interview, 14 June 2010.

13 Bruce Walker, recorded interview, 8 August 2008.

14 Ralph Folds, *Crossed Purposes: The Pintupi and Australia's Indigenous Policy*, University of NSW Press, Sydney, 2001, p. 62. Walker, 'Sustainability in Indigenous Australia', conference presentation, Murdoch University, 12–14 July 2006, p. 4.

15 Brian Singleton, recorded interview, 24 May 2010.

16 J.C. Altman, *Beyond Closing the Gap: Valuing Diversity in Indigenous Australia*, Centre for Aboriginal Economic Policy Research, Australian National University, Canberra, 2009, Discussion Paper No. 54/2009, p. 14.

17 Rose Kunoth-Monks, 'Land and Culture: Necessary but not Sufficient for the Future. Identity in the 21st Century', undated CAT presentation (c. 2006), Bruce Walker's personal papers.

18 Robyn Ellis, recorded interview, 13 June 2010.

19 Peter Taylor, recorded interview, 19 August 2008.

20 Andre Grant, recorded interview, 24 May 2010.

21 Bruce Walker, 'The Emperor's New Clothes', presentation at the Northern Territory Indigenous Housing Workshop, 11–12 July 2006. Bruce Walker, recorded interview, 22 September 2007. See Federal Race Discrimination Commissioner, *Water: A Report on the Provision of Water and Sanitation in Remote Aboriginal and Torres Strait Islander Communities*, Australian Government Publishing Service, Canberra, 1994, pp. 93–100, 171. Human Rights and Equal Opportunity Commission, *Review of the Water Report*, Human Rights and Equal Opportunity Commission, Sydney, 2001, p. 14.

22 See www.desertknowledge.com.au/remoteFOCUS. See especially Bruce Walker, Doug Porter, and Ian Marsh, *Fixing the Hole in Australia's Heartland: How Government Needs to Work in Remote Australia*, Desert Knowledge Australia/ remoteFOCUS, Alice Springs, 2012.

23 Chaney, Walker, Marsh, Porter, and Huigen, 'Submission to Senate Standing Committee on Community Affairs', p. 8.

24 Jon Altman, 'What Future for Remote Indigenous Australia? Economic Hybridity and the Neoliberal Turn', in Jon Altman and Melinda Hinkson (eds.), *Culture Crisis: Anthropology and Politics in Aboriginal Australia*, University of New South Wales Press, Sydney, 2010, p. 270.

25 *Closing the Gap: Prime Minister's Report 2012*, p. 3.

26 Robyn Grey-Gardner, recorded interview, 14 June 2010.

27 Mark Moran, recorded interview, 12 June 2010.

28 Bruce Walker, 'Science and Technology for Remote Communities', keynote address, Science and Technology for Remote Communities Conference, Murdoch University, July 1988, in CAT archives, box 2.

29 Su Groome, recorded interview, 25 May 2010.

30 www.icat.org.au

31 CAT Board meeting, 26 August 2002.

32 Jenny Kroker, recorded interview, 22 August 2008.

Index

Numbers in bold indicate illustration numbers in the illustration section

Wakefield Press is an independent publishing and
distribution company based in Adelaide, South Australia.
We love good stories and publish beautiful books.
To see our full range of books, please visit our website at
www.wakefieldpress.com.au
where all titles are available for purchase.

Find us!

Twitter: www.twitter.com/wakefieldpress
Facebook: www.facebook.com/wakefield.press
Instagram: instagram.com/wakefieldpress